Elements of Solid State Physics

Elements of Solid State Physics

SECOND EDITION

M. N. RUDDEN and **J. WILSON**
University of Northumbria at Newcastle
Newcastle upon Tyne, UK

JOHN WILEY & SONS

Chichester · New York · Brisbane · Toronto · Singapore

Other Wiley Editorial Offices

John Wiley & Sons, Inc., 605 Third Avenue,
New York, NY 10158-0012, USA

Jacaranda Wiley Ltd, G.P.O. Box 859, Brisbane,
Queensland 4001, Australia

John Wiley & Sons (Canada) Ltd, 22 Worcester Road,
Rexdale, Ontario M9W 1L1, Canada

John Wiley & Sons (SEA) Pte Ltd, 37 Jalan Pemimpin # 05-04,
Block B, Union Industrial Building, Singapore 2057

Library of Congress Cataloging-in-Publication Data
Rudden, M. N.
    Elements of solid state physics / M. N. Rudden and J. Wilson. —
2nd ed.
        p.   cm.
    Includes bibliographical references and index.
    ISBN 0 471 92972 7 (cloth): ISBN 0 471 92973 5 (pbk.)
    1. Solid state physics.   I. Wilson, J. (John), 1939–   .
II. Title.
QC176.R78   1992
530.4′1 — dc20   92-36621
                                                      CIP

British Library Cataloguing in Publication Data
A catalogue record for this book is
available from the British Library

ISBN 0 471 92972 7 (cloth)
ISBN 0 471 92973 5 (paper)

Phototypeset in 10/12pt Times from authors' disks by
Dobbie Typesetting Limited, Tavistock, Devon.
Printed and bound in Great Britain by Biddles Ltd, Guildford.

Contents

Preface to First Edition   ix

Preface to Second Edition   xi

Prologue   xiii

1 SOME ASPECTS OF MODERN PHYSICS   1
1.1 The formula for blackbody radiation   2
1.2 The specific heat of solids   6
1.3 The photoelectric effect   8
1.4 de Broglie's hypothesis   13
1.5 Heisenberg's uncertainty principle   17
1.6 Spectral lines   19
1.7 Atomic structure   20
1.8 The Bohr theory of the hydrogen atom   22
1.9 Summary   30
Problems   30

2 STRUCTURE OF CRYSTALLINE SOLIDS   33
2.1 Structure of solids   34
2.2 X-ray diffraction by crystals   35
   2.2.1 X-ray diffraction—the powder method   38
2.3 The geometry of perfect crystals   39
   2.3.1 Centre of symmetry   39
   2.3.2 Axes of symmetry   41
   2.3.3 Symmetry planes   41
   2.3.4 Rotational inversion symmetry   42
2.4 Bravais lattices   42
2.5 Miller indices   45
2.6 Crystal unit cell structures   47
   2.6.1 Cubic structures   48
   2.6.2 Hexagonal structure   49
2.7 Atomic bonds   51
   2.7.1 Ionic bond   53
   2.7.2 Covalent bond   55
   2.7.3 Metallic bond   57
   2.7.4 Other bonds   57

2.8   Real crystals   58
2.9   Defects in crystalline materials   58
      2.9.1 Point defects   58
      2.9.2 Line defects   59
      2.9.3 Plane defects   62
      2.9.4 Phonons   63
2.10 Crystal growing techniques   64
      2.10.1 The Czochralski method   66
      2.10.2 The Bridgman–Stockbarger method   66
      2.10.3 The floating-zone method   68
2.11 Device fabrication   68
2.12 Summary   76
Problems   76

3   THEORIES OF CONDUCTION AND MAGNETISM   79
3.1   Charge carriers in solids   80
3.2   Conduction in metals—classical approach   81
      3.2.1 Ohm's law   82
      3.2.2 Wiedemann–Franz law   85
3.3   Breakdown of the classical theory of conduction   87
      3.3.1 Mean free paths   87
      3.3.2 Specific heats   88
      3.3.3 The Hall effect   88
      3.3.4 Additive nature of resistivity   91
3.4   The magnetic properties of solids   92
      3.4.1 Atomic theory of magnetism   94
3.5   Classification of types of magnetism   99
      3.5.1 Diamagnetism   102
      3.5.2 Paramagnetism   103
      3.5.3 Ferromagnetism   103
      3.5.4 Ferrimagnetism   109
3.6   Superconductivity   111
3.7   Summary   113
Problems   114

4   ENERGY BANDS IN SOLIDS   117
4.1   The wave equation   117
      4.1.1 Stationary states   118
      4.1.2 Physical interpretation of Ψ 119
4.2   Solution of the Schrödinger equation for a particle in a
      one-dimensional box   121
4.3   Solution of the Schrödinger equation for a particle in a
      three-dimensional box   124

4.4   Removal of degeneracy   127
4.5   Aggregates of atoms   128
      4.5.1 Origin of energy bands   128
      4.5.2 The Kronig–Penney model   131
      4.5.3 The motion of electrons   134
4.6   The Fermi level   135
4.7   Experimental investigation of energy bands   138
4.8   Classical and quantum statistics   141
4.9   Classification of solids according to band theory   144
4.10 Summary   145
Problems   145

5   QUANTUM THEORY OF CONDUCTION   147
5.1 The behaviour of electrons in energy bands   147
    5.1.1 Filled bands   147
    5.1.2 Partially filled bands   148
    5.1.3 Positive holes   151
    5.1.4 Mean free paths   156
5.2 The electronic properties of semiconductors   158
    5.2.1 Intrinsic semiconductors   158
    5.2.2 Extrinsic semiconductors   159
5.3 The Fermi level in semiconductors   162
5.4 Conductivity of semiconductors   165
5.5 The optical properties of semiconductors   169
    5.5.1 The Haynes–Shockley experiment   172
5.6 Thermionic emission   178
5.7 Summary   180
Problems   181

6   SEMICONDUCTOR DEVICES   183
6.1 Metal–metal junctions   183
6.2 Metal–semiconductor junctions   189
6.3 Semiconductor–semiconductor junction devices   194
    6.3.1 The pn junction diode   194
    6.3.2 The pn junction without an applied bias   197
    6.3.3 The pn junction with an external applied bias   199
6.4 Reverse bias breakdown   203
    6.4.1 Zener breakdown   204
    6.4.2 Avalanche breakdown   204
6.5 The tunnel diode   207
6.6 Junction capacitance   210
6.7 Slope resistance r_e 211

6.8 The bipolar junction transistor   212
   6.8.1 The npn transistor   214
   6.8.2 The pnp transistor   215
   6.8.3 Transistor characteristics and gain parameters   216
   6.8.4 Switching mode   219
6.9 Field effect transistors   220
   6.9.1 The junction field effect transistor   221
   6.9.2 The metal oxide semiconductor transistor   223
6.10 Optoelectronic devices   226
   6.10.1 Photodetectors   226
   6.10.2 Photodiodes   228
   6.10.3 Light emission   231
6.11 Concluding remarks   238
Problems   239

Suggested Reading List   241

Discussion Questions   245

Answers to Problems   249

Appendices
  Appendix 1   Properties of Semiconducting Materials (at 300 K)   253
  Appendix 2   SI Units and Values of Physical Quantities   255
  Appendix 3   Motion of an Electron in a Periodic Lattice:
             The Kronig–Penney Model   257

Index   261

Preface to First Edition

Some years ago a discussion group was formed within the Department of Physics and Physical Electronics, Newcastle upon Tyne Polytechnic to consider the teaching of solid state physics at a level equivalent to the first year of HND and Degree courses. During these discussions it became apparent that there were no textbooks at an appropriate level. Both authors subsequently were required to teach solid state physics to students having at best the qualification of A-level General Certificate of Education in physics. The lack of suitable texts to supplement lectures to such students proved to be a major handicap. Most of the then available books on solid state physics, excellent though they were, relied very heavily on concepts completely alien to first year students and catered largely for final year honours and postgraduate students. As such, these books were and still are unsuitable for the evergrowing number of students studying solid state physics at an introductory level, many of whom require only sufficient background to understand the operation of modern solid state electronic devices.

It was with these considerations in mind that the authors wrote their first text *A Simplified Approach to Solid State Physics* (Butterworths 1971) which they hoped would serve as an introduction to the subject by introducing the concepts as simply as possible often with the aid of analogies and without recourse to difficult mathematics. Largely as a result of using that text for a number of different courses and the consequent student feedback the authors felt that some re-arrangement and clarification of the material would be desirable. The present book has therefore been written with these aims in mind. The authors have attempted further to elucidate some topics, present the material in what it is hoped will provide a more logical approach, and respond to changes in emphasis which have occurred since 1971. Problems have been introduced at the end of each chapter to help the readers' understanding of the material presented and to encourage them to become familiar with the book. The problems, many of which have been taken from examination papers set at Newcastle, are intended not just as a number-substituting exercise but also to extend the students' knowledge of the subject. Solutions, with hints, are given at the end of the book.

The authors are indebted to their colleagues for many fruitful and stimulating discussions and to their wives for their encouragement to undertake the task of rewriting the original text. They would like to thank Mrs Margaret Boardman for typing the manuscript and Mr A. Kitchen for preparing the electron micrographs of etch pits and the integrated circuit.

ix

Preface to Second Edition

It is now two decades since the authors wrote their first text *A Simplified Approach to Solid State Physics* as a response to the demand for an introduction to the theory of solid state devices, such as the diode and the junction transistor, without recourse to difficult mathematics, and relying as far as possible on analogies. Indeed, this grew out of the fact that the BSc Physical Electronics course at the then Newcastle Polytechnic had just abandoned thermionics in favour of solid state devices—a revolutionary step, since there were no appropriate first year level texts.

Ten years later the first edition of the present book appeared, in which an opportunity was taken to update and clarify the material in the light of developments in the curriculum and in the subject itself, hopefully to the advantage of first year degree students in physics or electronic engineering based courses. However, in the last decade there have been even greater changes in the school curriculum. Solid state electronics is a regular component of A-level physics courses, and indeed it is not uncommon to find GCSE pupils who are perfectly at ease with transistors, operational amplifiers and logic circuits, albeit at an operational level. Moreover, the next few years will see a dramatic reduction in the content of university physics degrees. More time will have to be allowed for students to assimilate material, and the initial mathematical content will be significantly reduced.

It is therefore timely to present the second edition of this book, in which an attempt has been made to include material that will possibly extend its use to second year level, and also to mention recent developments such as high temperature superconductors and optoelectronics. Furthermore, in response to demand from our own students we have also included a number of worked examples, which should serve to encourage those working independently to tackle the wider range of problems at the ends of each chapter, and encourage all readers to continue their study of solid state physics.

M. N. Rudden
J. Wilson

Newcastle upon Tyne, August 1992

Prologue

In recent years the expression 'solid state' has gradually become part of everyday speech, although it is doubtful whether many people would be aware of what these words mean. Indeed, in the history of physics the term is of fairly recent origin, and its first formal use was just before the start of World War II.

Interest in the behaviour of electrons in solids was mainly generated by R. W. Pohl, in the early twenties, and it was from his research work and that of his colleagues at Göttingen that the techniques for growing pure single crystals of rock salt eventually led to the building up of concepts which formed the basis of modern semiconductor physics. In 1933 Pohl predicted that valves in radios would eventually be replaced by small crystals in which the motion of electrons could be controlled. This prediction came true with the announcement in the *New York Times* on 1 July 1948 that a silicon device called a transistor had been used successfully as an amplifier. Thus began what has been termed the Electronic Revolution which has had an even more dramatic effect on society and everyday life than did its industrial counterpart in the nineteenth century. The discovery that more than one transistor could be put on to a single piece of silicon, once made, took on its own momentum and culminated in the integrated circuit — the now famous, or infamous, 'chip'. However, no matter how many devices may eventually be crowded on to a pin-head of silicon, the basic principles of solid state physics as outlined in this book still apply, and an understanding of the operation of the diode and transistor still provides the basis to appreciate the operation of the highly sophisticated solid state circuitry that is available today.

1 Some Aspects of Modern Physics

Most readers of this book will already be quite familiar with many aspects of what is now termed 'classical' physics. They should be conversant with Newton's laws of motion, the laws of geometrical optics, the nature of sound waves, the laws of electrostatics and of electromagnetism and a whole host of other phenomena.

This is precisely the situation in which physicists found themselves towards the end of the last century. All the well-established laws had been tested and proved over long periods, and apart from one or two controversies, such as whether light was composed of particle streams or whether it had a wave nature, everything seemed to fit neatly into place. There was no conceivable reason to doubt hypotheses that had been proposed by such great names as Newton, Hooke, Coulomb, Maxwell and Faraday.

However, a revolution was about to take place in the world of physics which would change the attitude of physicists for all time. The situation was very similar to that which existed in the world of music. Towards the end of the nineteenth century, people were quite familiar with the works of Bach, Beethoven and Mozart. They had been brought up in the tradition of 'classical' music. It is hardly surprising, therefore, that the early works of such composers as Stravinsky and Bartok seemed incomprehensible and an apparent contravention of all the well-established laws of rhythm, harmony and counterpoint. Nevertheless, 'modern' music has developed and is now appreciated by a wide range of listeners.

Why then was the 'physical revolution' so long in coming? The main reason was largely technical. Until the middle of the last century, experiments could deal only with large-scale phenomena such as the collisions of elastic spheres, the vibrations of waves on a string and free fall under gravity. Only when it became possible to study small-scale phenomena such as atomic collisions, the change of energy states of atoms and the interaction of light waves with matter did the inadequacies of classical theory become evident. Moreover, it had never occurred to anyone to look 'inside' atoms since they were believed to be structureless.

Once started, the revolution gathered momentum, increasingly refined experimental techniques were evolved, and more and more sophisticated theories were put forward to explain the experimental results. Thus, in the space of a few years, J. J. Thomson demonstrated the existence of the electron (the

1

fundamental unit of electricity), H. Bequerel discovered natural radioactivity and W. C. Röntgen discovered X-rays. Meanwhile, M. Planck put forward the hypothesis that radiant energy, in its interaction with matter, behaved as though it were quantized—that is as if it existed in very small discrete amounts. This led ultimately to the development of quantum mechanics. Also during this period, A. Einstein reconsidered the fundamental concepts of physics and was led to his theories of relativity.

Perhaps the most significant advance was the development of the quantum theory, which was able to explain several experimental observations that classical physics, based on the laws of I. Newton, could not.

A brief survey of some of the experiments and theories that laid the foundations of modern physics will now be given.

1.1 THE FORMULA FOR BLACKBODY RADIATION

Planck's quantum hypothesis arose out of the inability of classical physics to explain the observed distribution of energy in the continuous spectrum emitted by an incandescent body. The electromagnetic radiation emitted by the surface of a hot body is termed *thermal radiation*. The spectral distribution and amount of energy radiated depend largely on the temperature of the surface, and also on its nature, that is on its *emissivity* ϵ. Towards the end of the nineteenth century several people investigated the nature of thermal radiation. In particular they investigated *blackbody* radiation such as would be emitted by a perfect emitter ($\epsilon = 1$). A blackbody also absorbs all the radiation falling on it, that is it also has an *absorptivity* of unity. For practical purposes a blackbody can be realized by making a small aperture in an otherwise enclosed, hollow cavity, whose inner wall is blackened and maintained at a constant absolute temperature T. The radiation emitted from the aperture is then characteristic of the thermal radiation from a blackbody. The radiation inside the cavity interacts with the electrons, atoms and molecules in the cavity walls by reflection, scattering, absorption and emission until it is in thermal equilibrium with the cavity.

This radiation was studied by O. Lummer and E. Pringsheim (1899). They measured the power emitted by the aperture per unit area within a small wavelength range λ to $\lambda + d\lambda$. These results are shown in Figure 1.1, in which the *spectral radiance* (power emitted per unit area per unit wavelength interval), I_λ, is plotted as a function of wavelength λ, for a range of cavity temperatures. From Figure 1.1 it can be seen that as the temperature increases (a) the total emitted power increases, (b) the power emitted at a given wavelength increases and (c) the wavelength λ_{max}, corresponding to maximum power, moves to shorter wavelengths.

The first of these observations conforms with the Stefan–Boltzmann law (obtained empirically by J. Stefan in 1874 and derived using thermodynamics

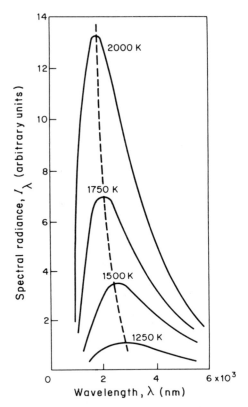

Figure 1.1. Distribution of energy in the spectrum of a blackbody radiator at various temperatures

by L. Boltzmann in 1884), which states that the total power per unit area, I, emitted by a blackbody is proportional to the fourth power of the absolute temperature,* that is

$$I = \sigma T^4 \tag{1.1}$$

where σ is the Stefan–Boltzmann constant ($\sigma = 5.67 \times 10^{-8} \, \text{W m}^{-2} \, \text{K}^{-4}$). The last of the above observations confirmed a prediction by W. Wien (1893) that λ_{max} would vary inversely with the absolute temperature, that is,

$$\lambda_{max} T = b \tag{1.2}$$

where b is a constant ($b = 2.898 \times 10^{-3} \, \text{mK}$). Equation (1.2) is known as Wien's displacement law.

_____

*The total power I is given by

$$I = \int_0^\infty I_\lambda \, d\lambda$$

These results agree with our everyday experience: a body radiates more when it is hot than when it is cold and the spectrum of a hot body has its peak at a shorter wavelength than the peak in the spectrum of a cooler body. We can observe these changes qualitatively by noting the change in the radiation, in both quantity and colour, of a piece of metal as its temperature is raised. Initially no visible radiation can be seen; then at about 500°C the metal glows a dull red and then progressively appears bright red, orange-red and eventually white hot as the temperature continues to rise. We note, however, that these colours do not correspond to the wavelength λ_{max} at the peak of the spectrum.

Example 1.1

Calculate the wavelength at the peak of the spectrum emitted by a blackbody at a temperature of 550°C. Also calculate the temperature of a blackbody that has a peak emission at a wavelength of 650 nm (red). Calculate the total power emitted in each case.

Solution. From equation (1.2) $\lambda_{max}T = 2.898 \times 10^{-3}$. Hence:

(a) At a temperature of 550°C, that is 823 K, we have

$$\lambda_{max} = \frac{2.898 \times 10^{-3}}{823} = 3.52 \ \mu m$$

which is in the near infra-red part of the spectrum.

(b) For a peak wavelength $\lambda_{max} = 650$ nm we have

$$T = \frac{2.898 \times 10^{-3}}{650 \times 10^{-9}} = 4458 \ K$$

(c) The total power per unit area in (a) is $5.67 \times 10^{-8} \times (823)^4$ or 2.6×10^4 W m^{-2} and in (b) is $5.67 \times 10^{-8} \times (4458)^4$ or 2.24×10^7 W m^{-2}.

All attempts to derive the distribution of radiated power shown in Figure 1.1 based on classical physics were unsuccessful, though an empirical formula derived by Wien, using thermodynamic arguments, gave good agreement with experiment for small values of T, in the wavelength range $\lambda < \lambda_{max}$.

Lord Rayleigh and J. Jeans using classical physics also attempted to correlate theory and experiment. They assumed that the radiation was emitted by electric oscillators, which generated waves that formed a set of standing waves, or modes, by reflection to and fro within the cavity. These can be visualized as a three-dimensional generalization of standing waves on a string. Rayleigh and Jeans then assumed that the principle of equipartition of energy applied so that the average energy associated with each mode of vibration would be kT, that is $\frac{1}{2}kT$ each for the potential energy and kinetic energy associated with the mode. They obtained the following relationship for the spectral radiance I_ν (power per unit area per unit frequency interval):

$$I_\nu = \frac{2\pi\nu^2 kT}{c^2} \tag{1.3a}$$

where ν is the frequency. Equation (1.3a) is often expressed in wavelength terms, namely

$$I_\lambda = \frac{2\pi ckT}{\lambda^4} \tag{1.3b}$$

Equations (1.3a) and (1.3b) are statements of the Rayleigh–Jeans formula. Equation (1.3b), for example, gives reasonable agreement with experiment at long wavelengths. It is totally invalid at short wavelengths, however, where it predicts that more and more radiation will be emitted as the wavelength decreases, in complete contradiction to the observations. This, the so-called ultraviolet catastrophe, suggested that there was a fundamental error in the classical approach.

The correct blackbody radiation formula was derived by Planck (1901) who, like Rayleigh and Jeans, supposed that the emission and absorption of radiation was due to electric oscillators. Planck, however, did not accept that the equipartition principle should apply, but rather assumed that the energy associated with each mode of oscillation was quantized; that is the energy could only exist in integral multiples of some smallest amount, which he called a *quantum*. He further postulated that the energy of the quantum was proportional to the frequency of the radiation. Thus the energy could not vary continuously but had to take one of the discrete values 0, $h\nu$, $2h\nu$, $3h\nu$ . . ., where h (Planck's constant) was introduced as the constant of proportionality between energy and frequency. Consequently, energy can only be emitted or absorbed by an oscillator in units of $h\nu$, and the emitted energy is in the form of a quantum of electromagnetic radiation or a photon.

From these assumptions Planck was able to derive a formula in complete agreement with the experimental curves shown in Figure 1.1. Planck's formula* is

$$I_\nu = \frac{2\pi h\nu^3}{c^3}\left(\frac{1}{\exp[(h\nu/kT)-1]}\right) \tag{1.4}$$

---

*The formula quoted here is for the spectral radiance I_ν, that is the power radiated per unit frequency interval from a small aperture in a blackbody cavity. Many texts quote the formula for the radiation density inside the cavity, u_ν, that is the energy per unit volume per unit frequency interval. We note that $I_\nu = u_\nu c/4$ is more directly comparable with the experimental data.

It is interesting to note that both Wien's law and the Stefan–Boltzmann law can be derived fairly easily from the Planck radiation formula (see Problem 1.2).

It must be emphasized that the concept of a discrete set of energy values or energy levels for a system emitting radiation is completely contrary to classical ideas. Classically, energy changes are smooth and the total energy can have continuous values (corresponding to $h = 0$), while in Planck's hypothesis energy changes are abrupt and discontinuous, and the total energy can assume only discrete values. The derivation of Planck's formula represented the initial step in the development of what came to be called modern physics. The phenomena discussed briefly in the following sections reinforce Planck's concept of quantization and the need to use quantum physics to describe them.

1.2   THE SPECIFIC HEAT OF SOLIDS

The inadequacy of classical physics was further demonstrated when experimental techniques developed so that it became possible to measure the variation of the specific heat of solids over a wide temperature range and it was shown that the specific heat tended towards zero at very low temperatures, as shown in Figure 1.2.

Dulong and Petit (1819) found empirically that the product of atomic weight and specific heat capacity* is the same for all solid substances. At ordinary temperatures, this 'law' holds for many materials — the average value of the product, called the atomic heat (or molar heat capacity), being $25.7 \, \text{J mol}^{-1} \text{K}^{-1}$.

However, there are a few notable exceptions, including boron, silicon and carbon in diamond form, the molar heat capacities of which at room temperature are 14.0, 20.7, and $6.1 \, \text{J mol}^{-1} \text{K}^{-1}$ respectively. These materials were the subject of much research, and their molar heat capacities were soon found to vary rapidly with temperature. Indeed, it was noted that the molar heat capacity of diamond increased twofold between 300 and 500 K. The most important discoveries, however, were that the molar heat capacities of all solids decreased rapidly at sufficiently low temperatures, and that the curves were all of the same form and could be brought almost into coincidence by suitable adjustment of the temperature scale. Such a striking regularity, one would suspect, would have its origin in a simple general principle.

Classically, the atoms of a solid merely vibrate about certain mean equilibrium positions, and do not wander about as do the atoms of a liquid or gas. Thus, they behave rather like three-dimensional simple harmonic oscillators with three degrees of freedom. As these oscillators are bound to the mean positions, they

*The term specific heat is no longer recommended, current practice being to use the specific heat capacity or molar heat capacity.

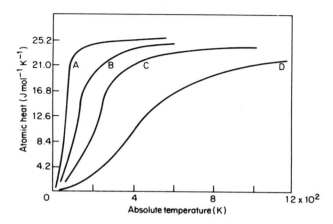

Figure 1.2. Variation of atomic heat with temperature. A, lead; B, aluminium; C, silicon; D, diamond

have both kinetic and potential energy, and by the law of equipartition of energy, the total internal energy E in a mole of any solid is

$$E = N \times \frac{6kT}{2} \qquad (1.5)$$

where N is Avogadro's number (the number of atoms in a gram molecular weight, or mole) and k is Boltzmann's constant. The molar heat capacity is thus

$$C_v = \frac{dE}{dT} = 3Nk \qquad (1.6)$$

or $C_v = 3R$ as $k = R/N$ (where R is the universal gas constant, equal to 8.32 J mol^{-1} K^{-1}). This expression for the molar heat capacity derived purely on a classical basis is independent of temperature, which is clearly at variance with the experimental observations, although the predicted value for $C_v = 3R = 25$ J mol^{-1} K^{-1} is in good agreement with the Dulong and Petit law. Einstein (1907) improved the agreement between experiment and theory by replacing the classical value kT for the mean energy of an oscillator by the value $h\nu/[\exp(h\nu/kT) - 1]$ which had been derived from Planck's quantum theory. This gave

$$C_v = 3R \left\{ \frac{\exp(h\nu/kT)}{[\exp(h\nu/kT) - 1]^2} \left(\frac{h\nu}{kT} \right)^2 \right\} \qquad (1.7)$$

which at high temperatures approached $3R$, but decreased rapidly at low temperatures. The fit with experiment is good except at very low temperatures, but this discrepancy was resolved by P. Debye in 1912.

In Einstein's model of solids, each atom is regarded as vibrating independently of its neighbours. Debye, however, assumed that solids behave as continuous elastic bodies. The internal energy, rather than being due to the vibrations of the individual atoms, is assumed to reside in elastic standing waves. These waves, like the electromagnetic waves in a cavity, have quantized energy equal to $h\nu$ (as do photons). A quantum of vibrational energy in a solid is called a *phonon* (we shall return to this concept in Chapter 2 of the text). Phonons travel at the speed of sound in the solid since sound waves are elastic in nature. Debye then assumed that the average energy of a standing wave was the same as that of Einstein's oscillators but, unlike Einstein, that there was a range of allowed frequencies up to some maximum frequency ν_m. The value of ν_m is related to the number of atoms per unit volume in the solid (put another way, it would be meaningless to consider the propagation of an elastic wave in a solid whose wavelength is smaller than the interatomic separation). Although not universally applicable, Debye's model does lead to a very good fit between experiment and theory.

It is noteworthy that if Planck's constant h were to become exceedingly small, that is the quantum of energy were to approach zero so that energy changes became continuous, then we could write the mean energy of an oscillator given above as

$$\bar{E} = h\nu \left[1 + \frac{h\nu}{kT} + \left(\frac{h\nu}{kT} \right)^2 + \cdots - 1 \right]^{-1}$$

or, as $h\nu \ll kT$,

$$\bar{E} = h\nu \left(\frac{h\nu}{kT} \right)^{-1} = kT$$

as in classical equipartition of energy. Similarly we note that equation (1.7) would give $C_v = 3R$.

1.3   THE PHOTOELECTRIC EFFECT

The work of J. C. Maxwell (1864) predicted the existence of electromagnetic waves; these were discovered experimentally by H. Hertz (1887). In the course of his work, Hertz discovered that a spark gap could discharge more readily if it was illuminated by ultraviolet light. Subsequently, W. Hallwachs showed that ultraviolet light falling upon a negatively charged insulated plate caused the negative charge to leak away. If the plate was positively charged, however, no such effect was observed. An explanation of these observations was provided by the discovery of the electron. It was suggested that electrons are liberated from the plate by the ultraviolet light. This was confirmed by e/m measurements for the 'photoelectric rays', the value being the same as that measured by Thomson for electrons.

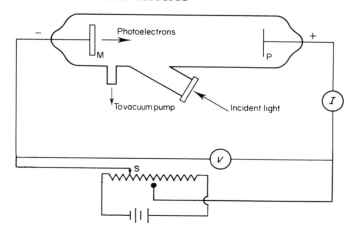

Figure 1.3. Simple apparatus used for investigating the photoelectric effect

An experimental arrangement, such as that used by P. Lenard (1902), for investigating the photoelectric effect is shown in Figure 1.3. The sign and magnitude of the potential difference between the electrode P and the metal plate M can be varied by moving the sliding contact S along the resistor. Thus, when light falling on to the metal plate causes the photoelectrons to be ejected, they will either be attracted or repelled by P according to whether P is positive or negative with respect to M.

The relationship between the photocurrent I resulting from the collecting of the photoelectrons, the potential difference V between M and P, and the irradiance of the illumination is shown in Figure 1.4. Two main points emerge. Firstly, with steady irradiance the photocurrent increases rapidly as the potential difference increases (for P positive with respect to M) and quickly reaches a fixed maximum value. The maximum value of photocurrent is directly proportional to the light irradiance. These observations indicate that all of the photoelectrons ejected are collected for quite small potential differences, and that the number of photoelectrons ejected is proportional to the light irradiance. Secondly, if the potential between M and P is reversed, the photocurrent is reduced because the electrons are repelled by P. At a certain 'stopping' potential V_0, independent of the light intensity causing emission, the photocurrent becomes zero, indicating that the photoelectrons are emitted with a maximum kinetic energy given by*

$$eV_0 = \tfrac{1}{2}mv^2_{max} \qquad (1.8)$$

---

*Throughout the text we have used e to indicate the *magnitude* of the electron charge. The negative *sign* of the charge will be included when the value of e is substituted into equations.

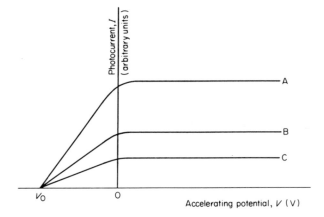

Figure 1.4. Photoelectric current I as a function of accelerating voltage V for various light irradiances falling on a given metal, and at fixed frequency of the light. A, full irradiance of the light; B, half irradiance; C, quarter irradiance

where v_{max} is the maximum velocity of emission of the photoelectrons. Those electrons with velocity of emission less than v_{max} are returned to the metal plate by lower voltages than the stopping potential.

The dependence of the stopping potential on the wavelength of the light was subsequently investigated by R. M. Millikan (1916), who prepared clean uncontaminated metal surfaces inside a vacuum chamber to eliminate the effects of surface dirt and oxide layers. Millikan's results for a range of metals are summarized graphically in Figure 1.5.

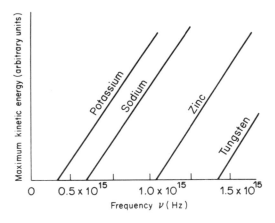

Figure 1.5. Maximum kinetic energy of the photoelectrons emitted from various metal surfaces plotted as a function of frequency

The straight line graphs of Figure 1.5 can be expressed by the equation

$$\tfrac{1}{2}mv_{max}^2 = eV_0 = h(\nu - \nu_0)$$

or

$$\tfrac{1}{2}mv_{max}^2 = eV_0 = h\nu - \phi \tag{1.9}$$

where h and $\phi = h\nu_0$ can be determined from the graphs. Two important points arise: the straight lines for the different metals all have the same slope, in other words h appears to be a universal constant (in fact, Planck's constant), whereas ϕ, the intercept on the energy axis, is a constant characteristic of the metal, called the *work function*. We note that ν_0 is the smallest frequency that can cause emission from a given surface; ν_0 is often called the *threshold frequency*.

Example 1.2

When a copper surface is illuminated with light having a wavelength of 254 nm it is found that the stopping potential is 0.24 V. Calculate the maximum velocity of the photoelectrons, and also the work function and threshold frequency for copper.

Solution. From equation (1.9) we have

$$v_{max}^2 = \frac{2eV_0}{m} = \frac{2 \times 1.6 \times 10^{-19} \times 0.24}{9.1 \times 10^{-31}}$$

that is

$$v_{max} = 2.9 \times 10^5 \text{ m s}^{-1}$$

Also from equation (1.9) we have

$$\phi = h\nu_0 = h\nu - eV_0$$

or

$$\phi = \frac{hc}{\lambda} - eV_0 = \frac{6.63 \times 10^{-34} \times 3 \times 10^8}{254 \times 10^{-9}} - 1.6 \times 10^{-19} \times 0.24$$

that is

$$\phi = 7.44 \times 10^{-19} \text{ J}$$
$$= 4.64 \text{ eV*}$$

The threshold frequency is given by

$$\nu_0 = \frac{\phi}{h} = \frac{7.44 \times 10^{-19}}{6.63 \times 10^{-34}}$$

*Energy on the atomic scale is often expressed in terms of electronvolts (eV), though this is not strictly a unit but the product of the electron charge e and a voltage V, $1 \text{ eV} = 1.6 \times 10^{-19}$ J.

that is

$$\nu_0 = 1.12 \times 10^{15}\,\text{Hz}$$

Attempts to explain the photoelectric effect in terms of classical physics met with failure, just as for other phenomena associated with the atom. The controversy as to whether light was of a wave or corpuscular nature had seemingly been settled in favour of waves by Young's and Fresnel's discovery of interference and diffraction effects. Maxwell later showed light waves to consist of electromagnetic waves similar in nature to those discovered by Hertz. It was supposed initially that the electric and magnetic fields associated with the light waves exerted forces on the electrons in the metals, liberating them from the metal surface. If this were so, one would expect that light of high irradiance with large electric and magnetic fields would give a high kinetic energy to the liberated electrons. This notion is contrary to experiment, which shows that the kinetic energy of the photoelectrons has a definite maximum independent of irradiance, as shown in Figure 1.4. Furthermore, it might be expected that high-irradiance low-frequency light (infra-red) should be at least as successful as ultraviolet light of low irradiance in liberating photoelectrons. Again, this supposition is not confirmed by experiment, which shows that light below the threshold frequency ν_0 is ineffective (Figure 1.5).

The explanation of the early observations was given by Einstein (1905), his theory later being confirmed by Millikan's measurements. Einstein based his explanation on the quantum theory of Planck: he assumed that radiant energy is not continuously distributed over a wavefront, as, for example, is the energy of water waves, but is concentrated in quanta or photons of energy $h\nu$ for light of frequency ν. Furthermore, Einstein assumed that each photon is so concentrated that its whole energy $h\nu$ may be transferred to one electron within the solid. Applying the principle of conservation of energy to the transfer, Einstein assumed that the energy of the photon split into two parts, one part being used to enable the electron to overcome the forces binding it to the metal, the other part appearing as the kinetic energy of the escaping electron. Hence the energy $h\nu$ of one photon is related to the kinetic energy $\frac{1}{2}mv^2$ of the escaping electron by the equation

$$h\nu = \phi + \tfrac{1}{2}mv^2 \tag{1.10}$$

This is the same as the empirical equation (1.9); thus quantum theory predicts an equation that is in agreement with experiment. The value of h determined photoelectrically (6.58×10^{-34} J s) agreed very closely with that determined by Planck from the blackbody energy distribution (6.53×10^{-34} J s). It should be noted that the units of Planck's constant are those of angular momentum, so that h can be thought of as the smallest possible amount of angular momentum. This energy transfer does not take place for all quanta incident on the surface, for many are reflected, scattered or absorbed by the surface without a photoelectron being emitted.

In Einstein's theory, light of high irradiance is described as being due to many quanta per second travelling as a ray. Thus intense light incident on a surface is expected to liberate many electrons and give large photocurrents, without giving a higher energy to any individual electron than does less intense light of the same frequency. This observation is verified by experiment. The fact that not all electrons leave the surface with the maximum kinetic energy $\frac{1}{2}mv_{max}^2$ is easily explained by assuming that not all electrons are equally firmly bound. Nor do all electrons escape along paths perpendicular to the surface, so only a vanishingly small number of electrons reach the electrode P when the theoretical stopping potential is applied. (Photoelectric emission from metal surfaces is dealt with further in Chapter 4.)

Although Einstein's theory gave an accurate description of the photoelectric effect, his ideas of light consisting of quanta were completely at variance with classical electromagnetic wave theory, which cannot explain the photoelectric effect. On the other hand, it is impossible to explain interference and diffraction in terms of 'corpuscles'. It must, therefore, be accepted that light in its interaction with light displays a different nature from light in its interaction with matter. There is a *dual* nature to light.

1.4   DE BROGLIE'S HYPOTHESIS

In 1924, L. de Broglie extended this dualism to suggest that 'particles of matter, and in particular electrons, may display particle and wave natures'. The first experimental support for de Broglie's hypothesis came in 1927 from the electron diffraction experiment of G. Davisson and L. H. Germer, shown in Figure 1.6. Electrons from the heated filament F were accelerated by a small potential difference and struck a nickel single crystal normal to one of its faces.

The intensity of electrons was measured at various angles of scattering for a range of accelerating voltages between 40 and 68 V. The scattered electrons did not show the diffuse pattern, which might be expected, but a well-defined series of maxima and minima resembling a diffraction pattern. The maxima were explained by assuming that in the directions in which they lie there is constructive interference—that is reinforcement of the electron waves scattered from the regularly spaced atoms in the nickel single crystal. This is a similar effect to X-ray diffraction, which is described in more detail in Chapter 2.

More striking electron diffraction patterns were produced by G. P. Thomson and P. Reid (1928) by passing a beam of high-energy cathode rays through thin metal foils. The cathode rays were produced in a discharge tube (Figure 1.7) operating at potentials up to 60 kV. After passing through the foil, the cathode rays were received on a photographic plate. The pattern on the photographic

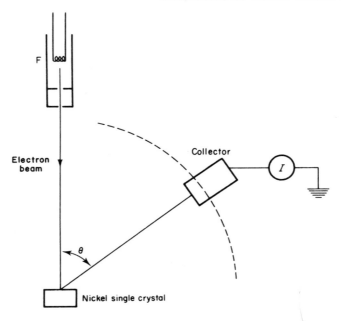

Figure 1.6. Diagram showing the arrangement used by Davisson and Germer to investigate electron diffraction. The collector is rotated along the dotted line about the point of incidence of the electron beam on the crystal

plate consisted of a series of well-defined concentric rings about a central spot. This kind of pattern, produced as it is by the randomly oriented microscopic crystals in the thin metal foil, is very similar in appearance to X-ray powder diffraction patterns. If the geometry of the crystals used in these experiments is known, the wavelength associated with the incident electrons can be determined from Bragg's law (see page 36).

Electron diffraction experiments showed that the wave associated with an electron has a wavelength given by

$$\lambda = \frac{h}{p} = \frac{h}{m\upsilon} \qquad (1.11)$$

where $p = m\upsilon$ is the momentum of the electron. For photoelectrons this formula can be derived fairly easily by means of equations from Einstein's theory of relativity. These equations, though not difficult, will not be introduced in this text.

For the Davisson–Germer experiment, the wavelength of the 'de Broglie' electron waves was derived from equation (1.11) and also from Bragg's diffraction law; the two results were in very good agreement, namely 1.66×10^{-10} and 1.65×10^{-10} m respectively for electrons accelerated by a

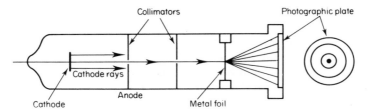

Figure 1.7. The experimental arrangement used by Thomson and Reid to investigate the diffraction of cathode rays (electrons) by thin metal foils

potential of 54 V (the potential at which the most clearly defined maxima and minima were observed).

Following the success of the application of de Broglie's hypothesis to electrons, attempts were made to demonstrate it for other particles. The diffraction of atoms and molecules (for example helium and hydrogen) was demonstrated by Stern and his colleagues in 1939. The diffraction of neutrons was also clearly established, and indeed electron and neutron diffraction techniques are now used extensively in scientific research and in industrial applications as a complementary 'tool' to X-ray investigations.

Example 1.3

Calculate the de Broglie wavelength of helium atoms heated to a temperature of 400 K.

Solution. The average energy of the atoms is given by

$$\tfrac{1}{2} M_{He} \bar{v}^2 = \tfrac{3}{2} kT$$

The mass of He atoms is 6.64×10^{-27} kg; hence their root mean square velocity is

$$v_{rms} = \sqrt{\bar{v}^2} = 1.58 \times 10^3 \text{ m s}^{-1}$$

The de Broglie wavelength of He atoms with this speed is

$$\lambda = \frac{h}{M_{He} v_{rms}} = 6.32 \times 10^{-11} \text{ m}$$

This value can be confirmed by diffraction experiments.

When Einstein first heard of de Broglie's hypothesis he remarked that the idea was probably sufficiently crazy to be correct, and indeed it does seem extreme to try and extend the idea to objects of everyday size. For example, does it mean that diffraction could be observed with a football? From the theory of diffraction the angle of deviation from the original direction of propagation of the waves is given by $\sin\theta = \lambda/d$, where d is the size of the diffracting object. Suppose that a 0.5 kg ball moving at $3\,\mathrm{m\,s^{-1}}$ passes through a slit of width 0.1 m. The de Broglie wavelength from equation (1.11) is $4.4 \times 10^{-34}\,\mathrm{m}$, which gives a diffraction angle of 4.4×10^{-33} radians. This means that to observe diffraction the screen would have to be placed a distance of the order of the size of the universe away! Thus, all moving objects exhibit wave nature, but the wavelength is usually undetectably small; it is only when the objects are of atomic dimension that the wavelengths become detectable.

A number of questions immediately arise. These include: What is the quantity which is varying in the de Broglie waves? What does the de Broglie wavelength represent? What is the velocity of propagation of the de Broglie waves? Space does not permit the full development of the answers to these and related questions. Suffice it to say that a particle, which we visualize as being localized in space, cannot be well represented by a wave equation of the form

$$y = a\sin(\omega t - kx) \tag{1.12}$$

which is infinite in extent. Instead, the particle is represented by a *group* of a large number of such waves, each of which has a slightly different frequency and velocity from the other members of the group. The waves within the group interfere with one another to give a localized resultant, which itself moves through space with a group velocity v_g (Figure 1.8a). The group velocity is different from the wave velocities v_w of the individual waves in the group; that is we have, from equation (1.12),

$$v_w = \frac{\omega}{k} \tag{1.13}$$

while

$$v_g = \frac{d\omega}{dk} \tag{1.14}$$

It is the wave group that represents the particle; indeed, it can be shown that the group velocity and particle velocity are the same (see Problem 1.7a).

An appreciation of some of these points can be obtained by considering the familiar phenomenon of beats. These are formed when two waves of closely similar frequencies interfere. The resultant (Figure 1.8b), which has the beat frequency, is also a travelling wave and represents the group formed by two waves (see Problem 1.7b). As with electromagnetic waves, the wave and particle aspects of moving bodies can never be observed simultaneously so we cannot say which is the correct description. All we can say is that in some respects a moving body exhibits wave properties, while in others it exhibits particle properties. An interpretation of this duality was given by W. Heisenberg (1926), and is described in the next section.

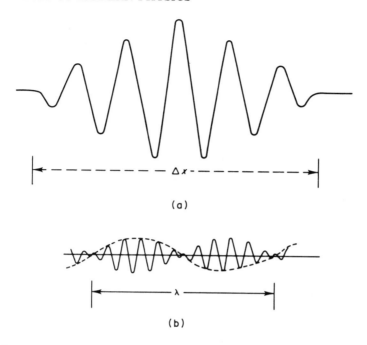

Figure 1.8. (a) The wave group formed by the interference of many waves of similar wavelengths. The narrower the group width Δx, the less uncertain is the position of the particle, but the greater is the uncertainty of the group wavelength. (b) Wave groups resulting from the interference of two waves of nearly equal wavelength. The group wavelength can be determined, but where is the particle?

1.5 HEISENBERG'S UNCERTAINTY PRINCIPLE

The fact that a particle can be visualized as a de Broglie wave group suggests that there is an uncertainty in the accuracy within which we can measure its properties. For example, referring to Figure 1.8(a) we see that, as the particle may be anywhere 'within' the group, we can only specify its position with an accuracy Δx. If the group were very narrow we could specify Δx very accurately, but we would be uncertain of the de Broglie wavelength. On the other hand, considering Figure 1.8(b) we can, in this case, estimate the wavelength accurately, but where is the particle located? Heisenberg extended this discussion to consider the uncertainties in the simultaneous measurement of the particle's position and momentum.

Consider, for example, the diffraction of a single photon as shown in Figure 1.9. When the photon arrives at slit S, its position along the y axis is uncertain and it could be anywhere across the slit width. If the uncertainty in its position

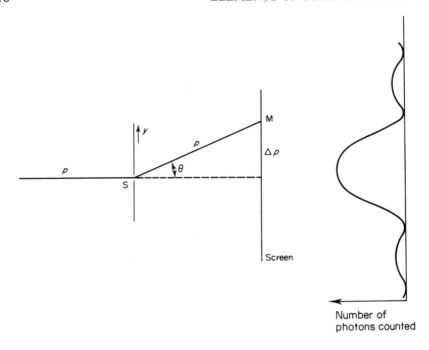

Figure 1.9. Photon diffraction at a slit

is Δy, then it is possible to equate Δy to d, the slit width. If the momentum of the photon before and after diffraction is p and the angle of deviation is θ, then it may be supposed that at the 'instant' of diffraction the photon receives a vertical impulse of uncertain magnitude, which causes the photon to deviate and contribute to a diffraction pattern produced on a screen by many such photons. The first minimum of the pattern is located at some position M. Clearly,

$$\Delta p_y = p \sin \theta = \frac{p\lambda}{d}$$

Therefore multiplying Δy and Δp_y to obtain the total uncertainty gives

$$\Delta y \Delta p_y = \frac{dp\lambda}{d}$$

Thus, finally,

$$\Delta y \Delta p_y \approx h \qquad\qquad (1.15)$$

where the equal sign has been replaced with a sign meaning 'of the order' since somewhat arbitrarily the first minimum was chosen as the typical diffraction angle; other 'derivations' give the right-hand side of equation (1.15) as $h/2\pi$. Equation (1.15) is the Heisenberg uncertainty principle, which states that it is impossible to know simultaneously the exact momentum (that is velocity) and position of even a single particle, and it formalizes the general philosophical principle that it is impossible ever to make an exact measurement; the very act of measurement upsets the situation that existed prior to making the measurement.

Example 1.4

Estimate how accurately the velocity of the electron in a hydrogen atom can be known.

Solution. We shall see below (Section 1.8) that the electrons in hydrogen atoms (in their ground state) have an orbital radius of about 10^{-10} m; hence

$$\Delta x \approx 10^{-10}$$

Using Heisenberg's uncertainty principle,

$$\Delta p = m\Delta v \approx \frac{h}{\Delta x}$$

$$\Delta v \approx \frac{h}{m \times 10^{-10}} \approx 7.6 \times 10^6 \text{ m s}^{-1}$$

Classical physics suggested a deterministic view of the Universe; once the positions and velocities of all moving objects were known at a given instant, their subsequent positions and velocities could be predicted accurately. The uncertainty principle says that this view can not hold.

The outcome of the notion of a wave–particle duality was therefore very profound, and from a mathematical point of view it was a natural step to seek a wave equation which could be used to describe matter waves and apply it to the problem of electrons in atoms. This is a major topic and is considered in Chapter 4. For the time being we will restrict ourselves to the development of ideas relating to the emission of spectral lines by atoms.

1.6 SPECTRAL LINES

A further apparent breakdown of classical physics arose when attempts were made to explain the characteristic discrete line spectra emitted by the elements in gaseous form. Detailed examination of these spectra indicated that various lines belonged together in different 'spectral series'. The separation between the lines in an individual series decreased with decreasing wavelengths and eventually converged at a definite series limit. As soon as dependable wavelength measurements became available, investigators sought for harmonic relations in the lines—reasoning from the analogy of overtones in acoustics. The search proved fruitless, but Ritz, Balmer and others succeeded in deriving simple empirical formulae giving relationships between the wave numbers $\bar{\nu} = 1/\lambda$ of the lines in a given series. For example, Balmer (1885) investigated the simplest of all spectra, namely that due to hydrogen, and represented all the then known lines by the formula

$$\bar{\nu} = \frac{1}{\lambda} = R_\mathrm{H}\left(\frac{1}{2^2} - \frac{1}{n_1^2}\right) \tag{1.16}$$

where $n_1 = 3,4,5, \ldots$, R_H is now known as the Rydberg constant and λ is the wavelength.

This formula was the key to the representation of spectral series of other elements by similar relationships involving the difference between two terms. However, all attempts to predict this formula classically, or indeed to explain the existence of discrete spectral lines, were unsuccessful. As will be seen in Section 1.8, it was left to Bohr, using the concept of quantized energy, to fit theory to experiment.

At this stage, it should be pointed out that, although Planck's hypothesis laid the foundations of modern quantum theory and wave mechanics, his original ideas have since been modified considerably. For the purposes of this text, his two fundamental postulates which have been retained permanently are:

(a) An oscillator, or any similar physical system, has a discrete set of possible energy values or levels (energies intermediate between these allowed values never occur).
(b) The emission or absorption of radiation (energy) is associated with transitions or jumps between two of these levels, the energy lost or gained by the oscillator being emitted or absorbed respectively as a quantum of radiant energy of magnitude $h\nu$.

1.7   ATOMIC STRUCTURE

The discovery of the electron by Thomson (1897) during his investigations on electric discharges in rarefied gases led to renewed interest in the problem of determining the structure of atoms. Thomson's experiments indicated that atoms must contain both positive and negative charge and that the properties of the negative charge are independent of the gas in the discharge tube. Deflection of cathode rays (electrons) by electric and magnetic fields enables the ratio of the electron charge e to the electron mass m to be determined; for cathode rays, this ratio was negative and independent of the gas. This indicated that electrons, as they were later called, were present in all atoms. Significantly, this e/m value was almost identical to that found by P. Zeeman (1896) (for the particles taking part in light emission) in his experiments on the effects of magnetic fields on spectral lines (see section 4.4), so it was then generally accepted that electrons were responsible for the emission of radiation from atoms.

Two questions arose: How many electrons are there in an atom and how are the electrons and positive charges arranged? Evidence from X-ray scattering investigations suggested that the number of electrons was about half the atomic weight. As far as their arrangement was concerned, it seemed on the basis of classical ideas that two conditions should be satisfied:

(a) The assembly of positive charge and electrons should be stable (that is the electrons should be fixed at specific equilibrium positions about which they could vibrate when disturbed or excited) and the vibrational frequencies should be such that the characteristic line spectra of the elements could be explained.
(b) Except when disturbed the electrons should remain at rest — otherwise they would radiate in accordance with the requirements of Maxwell's electromagnetic theory, which shows that any accelerated charge should emit radiation.

Several atomic models were proposed, but it was not until Rutherford and his co-workers carried out experiments on α-particle scattering from thin metal foils that a successful one emerged. The basis of the experiments, which were performed by Geiger and Marsden (1911), is schematically shown in Figure 1.10.

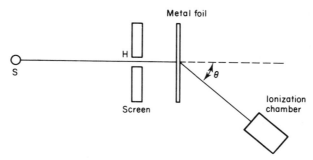

Figure 1.10. Sketch of the apparatus used by Rutherford to investigate the scattering of α-particles by metal foils

The positively charged α-particles emitted from the naturally radioactive source S, such as radon, were restricted to a narrow pencil by a hole H in the screen and fell on to the thin metal foil. The scattered α-particles struck a zinc sulphide screen, and the visible flashes of light, or scintillations, produced were viewed by a microscope (this detection system was later replaced by an ionization chamber). It was found that almost all of the α-particles passed straight through the foil and were deflected only slightly. A few of them, however, were scattered so strongly that they emerged on the same side of the foil as that on which they entered (that is the scattering angle $\theta > 90°$). The number of such α-particles increased as the foil thickness increased, up to a certain point. To explain these observations, Rutherford suggested that in an atom the positive charge and most of the mass is concentrated in a very small central region, now called the nucleus, about which the electrons are grouped in some configuration. Since atoms as a whole are neutral, the charge on the nucleus must be Ze, where e is the numerical electron charge and Z is the number of electrons in the atom (called the atomic number).

The existence of a stable configuration of positive and negative charges at rest presented an apparent difficulty (in fact, it can be shown that such an arrangement is impossible). One suggestion was that the electrons might orbit the nucleus in a similar way to that in which the planets orbit the sun. The objection to this is that, as mentioned above, electromagnetic theory would require the electrons to radiate energy as a result of the constant inward acceleration. Since this energy could only come from the system, the electron would spiral in towards the nucleus and, as will be shown in Section 1.8, would emit radiation of constantly increasing frequency, that is a continuous spectrum. This, of course, is contrary to the fixed-frequency spectral lines which are observed.

Bohr extended Planck's hypothesis in an attempt to explain the observed spectral lines emitted by hydrogen atoms.

1.8   THE BOHR THEORY OF THE HYDROGEN ATOM

The theory proposed to account for the spectrum of hydrogen by Niels Bohr in 1913 is of great historical significance, in that it provided a model for the behaviour of electrons in atoms. Starting with the Rutherford model of the hydrogen atom, Bohr set out to derive the energies which such an atom could have and to obtain a theoretical value for the Rydberg constant. The idea of the electron moving around the nucleus is very attractive; the similarity to planetary motion is apparent and the possibility of using the usual dynamic equations and the laws of electrostatics has an appeal. However, Bohr was faced with immediate difficulties and contradictions, and to make his theory work, he made what were then and still are unacceptable assumptions.

The central feature of the Bohr theory is the calculation of the gross structure of the spectrum of hydrogen, and his assumptions can be listed as follows:

(a) The Rutherford model consists of an electron orbiting the positively charged nucleus.
(b) The electron orbits can be exactly defined using the ordinary laws of classical mechanics.

(c) The nucleus and the electron are mutually attracted according to Coulomb's inverse square law.
(d) Since the electron possesses charge and is centripetally accelerated it should radiate energy and spiral into the nucleus; since this does not occur the normal laws of electromagnetic radiation must not apply in the permitted orbits.
(e) Not all the orbits permissible by classical mechanics are therefore possible; only those for which the total angular momentum is an integral multiple of Planck's constant divided by 2π are allowed, that is for which

$$mvr = \frac{nh}{2\pi} \tag{1.17}$$

(f) A transition between two orbits is possible by the emission or absorption of a single quantum of radiation. When an electron jumps from an orbit of energy E_1 to another of lower energy E_2 the energy difference is given by

$$E_1 - E_2 = h\nu \tag{1.18}$$

These latter two assumptions are usually called *Bohr's postulates*, and are completely alien to classical physics. In a sense they were made in order to make the theory fit the experimental observation of the discrete spectral lines. However, the theory does allow a very accurate formula for the overall structure of the hydrogen spectrum to be derived.

Let us assume that the orbital velocity of the electron is v in a circular orbit of radius r, as illustrated in Figure 1.11; then from Coulomb's law the force F upon the electron is given by

$$F = \frac{e^2}{4\pi\epsilon_0 r^2} \tag{1.19}$$

where ϵ_0 is the permittivity of free space. This inwardly directed force gives rise to an acceleration, so according to Newton's second law

$$F = \frac{e^2}{4\pi\epsilon_0 r^2} = \frac{mv^2}{r} \tag{1.20}$$

From equation (1.20), the kinetic energy E_K of the electron is given by

$$E_K = \frac{1}{2}mv^2 = \frac{e^2}{8\pi\epsilon_0 r} \tag{1.21}$$

Since the potential energy E_P of the electron is given by the equation

$$E_P = -\frac{e^2}{4\pi\epsilon_0 r} \tag{1.22}$$

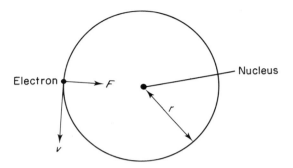

Figure 1.11. The Rutherford–Bohr model of the hydrogen atom, in which an electron of mass m and charge -1.6×10^{-19} C moves in a circular orbit around the nucleus, which has a charge of $+1.6 \times 10^{-19}$ C

the total energy E (kinetic plus potential) is given, from equations (1.21) and (1.22), by

$$E = -\frac{e^2}{8\pi\epsilon_0 r} \qquad (1.23)$$

The total energy of an electron within an atom is always negative since the potential energy is defined as zero for an ionized atom when the electron has moved to an infinite distance from the nucleus. Negative energy therefore corresponds to a bound state, and E is the binding energy. Bohr's quantum postulates can now be used in conjunction with the essentially classical approach given so far.

Eliminating v from equations (1.17) and (1.20) we see that the Bohr atomic radius is given by

$$r = \frac{n^2 h^2 \epsilon_0}{\pi m e^2} \qquad (1.24)$$

Example 1.5

Calculate the radius of the first Bohr orbit, that is when $n = 1$, and the corresponding electron velocity.

Solution. From equation (1.24) we have

$$r_1 = \frac{h^2 \epsilon_0}{m e^2 \pi} = 0.53 \times 10^{-10} \text{ m}$$

while from equation (1.20)

$$v_1 = \left(\frac{e^2}{4\pi m \epsilon_0 r_1}\right)^{1/2} = 2.19 \times 10^6 \text{ m s}^{-1}$$

The total energy E is given by equation (1.23) as

$$E = -\frac{e^2}{8\pi\epsilon_0 r} = -\frac{2\pi^2 m e^4}{n^2 h^2 (4\pi\epsilon_0)^2} \qquad (1.25)$$

Thus, from equation (1.18)

$$h\nu = E_1 - E_2 = \frac{2\pi^2 m e^4}{h^2 (4\pi\epsilon_0)^2}\left(\frac{1}{n_2^2} - \frac{1}{n_1^2}\right) \qquad (1.26)$$

where n_1 and n_2 are the values of the integer n corresponding to the energy levels E_1 and E_2 respectively. Hence the wave number $\bar{\nu} = 1/\lambda$ is given by

$$\bar{\nu} = \frac{\nu}{c} = \frac{2\pi^2 m e^4}{ch^3 (4\pi\epsilon_0)^2}\left(\frac{1}{n_2^2} - \frac{1}{n_1^2}\right) \qquad (1.27)$$

or

$$\bar{\nu} = R_{\mathrm{H}}\left(\frac{1}{n_2^2} - \frac{1}{n_1^2}\right) \qquad (1.28)$$

which is the Balmer formula (equation 1.16) if $n_2 = 2$ and $n_1 = 3,4,5, \ldots$. The Rydberg constant R_{H} can thus be calculated from atomic constants; the value obtained ($109.740 \times 10^5 \mathrm{\,m}^{-1}$) compares favourably with the experimentally determined value ($109.738 \times 10^5 \mathrm{\,m}^{-1}$).

The form of equation (1.28) suggests that other spectral series ought to exist for various values of n_2 — as in fact they do. Four other series have been observed: the Lyman series ($n_2 = 1$) in the ultraviolet, the Paschen ($n_2 = 3$), Bracket ($n_2 = 4$) and Pfund ($n_2 = 5$) series in the infra-red region of the spectrum. The way these series arise can be represented as shown in the energy level diagram (Figure 1.12) in which the various allowed energies for values of the so-called principal quantum number n from 1 to ∞ are shown.

Example 1.6

Calculate the energies corresponding to the first three values of the principal quantum number ($n = 1,2,3$), and hence calculate the wavelength emitted when an electron jumps from the energy level E_3 ($n = 3$) to level E_1 ($n = 1$).

Solution. From equation (1.25),

$$E_1 = \frac{-2\pi^2 m e^4}{h^2 (4\pi\epsilon_0)^2} = 2.182 \times 10^{-18} \mathrm{\,J}$$

or

$$E_1 = -13.62 \mathrm{\,eV}$$

E_1 is often called the ionization energy as it is the energy required to free the electron from the nucleus, that is to take it to an infinite distance.

Also, from equation (1.25),

$$E_n = \frac{E_1}{n^2}$$

Hence $E_2 = -3.40$ and $E_3 = -1.51$ eV. Therefore

$$E_1 - E_3 = 12.11 \text{ eV} = 1.94 \times 10^{-18} \text{ J} = h\nu = \frac{hc}{\lambda}$$

whence

$$\lambda_{3 \rightarrow 1} = 102.38 \text{ nm}$$

Let us pause at this point and note again the fundamental changes to experience that Bohr's postulates signified. As already suggested, as the electron has a centripetal acceleration it ought to emit electromagnetic waves, at a frequency which might be expected to be related to its orbital angular frequency $\omega = v/r$.

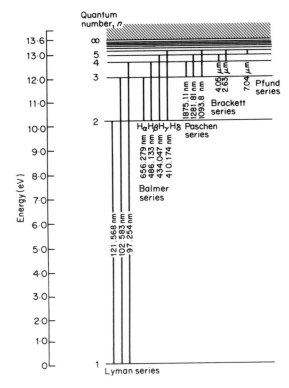

Figure 1.12. Energy level diagram for hydrogen, showing some of the principal lines of the spectrum

From equation (1.20) we see that

$$\omega^2 = \frac{v^2}{r^2} = \frac{e^2}{4\pi\epsilon_0 r^3} \tag{1.29}$$

while the total energy, from equation (1.23), is

$$E = \frac{-e^2}{8\pi\epsilon_0 r}$$

Now if the electron radiates electromagnetic waves it must lose energy; that is the total energy will become more negative, and hence the electron's orbital radius must, from equation (1.29), decrease. Consequently, its angular frequency, and the frequency of the emitted radiation, will increase continuously—in contradiction of the observations.

To resolve the contradiction Bohr postulated that only certain stationary orbits were allowed in which the electron does not emit energy. It is interesting that equation (1.17) expressing this postulate can be obtained by assuming that for an allowed orbit an integral number of electron de Broglie waves will fit exactly into the orbit (Figure 1.13a)—a circular standing wave; that is we have

$$n\lambda = \frac{nh}{p} = \frac{nh}{mv} = 2\pi r$$

or

$$mvr = \frac{nh}{2\pi}$$

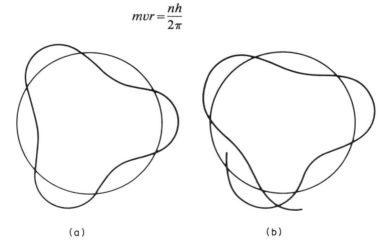

(a)   (b)

Figure 1.13. (a) The orbit of an electron in the hydrogen atom in which the electron de Broglie waves fit exactly into the circumference of the orbit. This corresponds to an allowed orbit. (b) In this case the de Broglie wavelength does not fit exactly into the circumference of the orbit. This is not an allowed orbit as the wave will interfere with itself destructively

The simple Bohr theory does give an adequate explanation of the spectral lines in hydrogen, but attempts to extend it to more complex atoms required further and further refinements. The fundamental weakness in the theory is that it is 'semi-classical' depending as it does on the concept of an electron as a particle moving in circular or (in a later refinement) elliptical orbits. This was a feature of what is termed the 'old' quantum theory; it might be regarded as an attempt to bridge the gap between the classical and quantum theories. Consequently, in the years between 1913 and the advent of wave, or quantum, mechanics in 1925 (see Chapter 4), refinements to the basic Bohr theory attempted to present a model of the atom in terms of a physical behaviour which could be visualized. To account for the appearance of the spectra of more complex atoms it was found necessary to introduce empirically four quantum numbers and to assign to them various restricted values. This description of electron behaviour in atoms is called the vector model of the atom and the four quantum numbers are:

Principal quantum number $n = 1, 2, 3, \ldots ,$
Orbital quantum number $l = 0, 1, 2, \ldots , (n-1)$
Magnetic quantum number $m_l = 0, \pm 1, \pm 2, \ldots , \pm l$
Spin quantum number $m_s = \pm \frac{1}{2}.$

The names reflect the attempt to associate classical behaviour with the electron. The principal quantum number determines the permitted values of the quantized electron energy. Similarly, the electron's orbital angular momentum is quantized, the magnitude of the allowed values being determined by l. Angular momentum is of course a vector and to describe it completely requires us to also specify its direction. This direction has greatest significance if an external magnetic field exists, whence the allowed orientations of the orbital angular momentum relative to the external field are specified by the magnetic quantum number m_l. Similarly, we often use the concept of electron spin, that is the electron spins as it orbits the nucleus (compare the spinning earth as it orbits the sun). The spin angular momentum has one value, but two orientations, that is parallel and antiparallel to the external magnetic field.

The important consequence of the introduction of these quantum numbers, albeit on an *ad hoc* basis, was that it was then possible to understand at last the systematic behaviour of the elements in the Periodic Table, constructed by Dimitri Mendeleev in 1868.

For every electron in an atom there is a unique set of four quantum numbers which completely specifies the state of that electron. They can be regarded as 'coordinates' for the electron, and one and only one electron can have a given set of coordinates. This is called the Pauli exclusion principle, from which it is possible to work out the electronic configuration of any atom. The application of the Pauli principle to the build-up of the elements of the Periodic Table is usually dealt with in detail in books on atomic physics. For the purposes of the present discussion, however, it is sufficient to know that, by systematically assigning sets of quantum numbers to successive electrons, a picture of the arrangement of electrons in atoms can be constructed. It is found that electronic shells and subshells exist which, when they contain a certain number of electrons, are considered to be full or closed. A shell exists for each value of n, and a subshell for each value of l associated with an n value. Thus for $n = 3$, l can have the three values 0, 1, and 2, and there are three subshells, each of which can accommodate $2(2l + 1)$ electrons. An s subshell ($l = 0$) can therefore contain a maximum of two electrons, a p subshell ($l = 1$) six electrons, a d subshell ($l = 2$) ten electrons, and so on.

On this basis it can be seen that the first shell ($n = 1$) can contain at most two electrons; this is designated the K shell. Similarly, an L shell ($n = 2$) can contain $2 + 6 = 8$ electrons, and an M shell can contain $2 + 6 + 10 = 18$ electrons. In general, therefore, a shell contains $2n^2$ electrons. The nomenclature for designating electronic shells is also largely historical and arises out of early optical and X-ray spectroscopic investigations. A summary of the electron configurations for the lighter elements is shown in Table 1.1. It can be seen from this table that all atoms consist of sets of closed shells outside of which are situated one or more valence electrons. Of particular interest are the elements helium and neon (and indeed other elements) which have completely filled shells and subshells. These are extremely stable since their 'closed' shells are reluctant to interact with other atoms.

The term valence is applied because it is these electrons that govern the chemical valency of the element, and more importantly, it is these electrons that

Table 1.1. The electron configurations of the lighter elements

Element	K shell 1s sub-shell	L shell 2s sub-shell	2p sub-shell	M shell 3s sub-shell	3p sub-shell	3d sub-shell	Electron configuration
Hydrogen	1						$1s^1$
Helium	2						$1s^2$
Lithium	2	1					$1s^2 2s^1$
Beryllium	2	2					$1s^2 2s^2$
Boron	2	2	1				$1s^2 2s^2 2p^1$
Carbon	2	2	2				$1s^2 2s^2 2p^2$
Nitrogen	2	2	3				$1s^2 2s^2 2p^3$
Oxygen	2	2	4				$1s^2 2s^2 2p^4$
Fluorine	2	2	5				$1s^2 2s^2 2p^5$
Neon	2	2	6				$1s^2 2s^2 2p^6$
Sodium	2	2	6	1			$1s^2 2s^2 2p^6 3s^1$
Magnesium	2	2	6	2			$1s^2 2s^2 2p^6 3s^2$
Aluminium	2	2	6	2	1		$1s^2 2s^2 2p^6 3s^2 3p^1$
Silicon	2	2	6	2	2		$1s^2 2s^2 2p^6 3s^2 3p^2$

are active in bonding mechanisms when atoms of the same element or dissimilar elements are brought together to form solids. The configurations shown represent the ground states of the atoms in question and provide a convenient shorthand notation to represent the atom, rather than drawing an energy level diagram. To write the electron configuration, the levels are written in order of increasing energy and the number of electrons in each level designated by a superscript. For example, the electron configuration of silicon is

$$1s^2\ 2s^2\ 2p^6\ 3s^2\ 3p^2$$

The ground state electron configuration of various elements will be used in Chapter 5. It is important to note that the apparently logical, systematic filling up of the electron shells and subshells does not continue for all 92 stable elements. It turns out that some apparently higher energy levels in fact have lower energies than apparently lower energy levels, and are therefore occupied by electrons first. Thus, for example, in the case of the important, so-called transition metals, which include iron, nickel, cobalt, copper and manganese, the 4s electron subshell is occupied before the 3d subshell is completely filled. This leads to some of the very interesting properties of this group of metals.

1.9 SUMMARY

The main purpose of this chapter was to introduce some new concepts which are essential for the description of solids and their physical properties. It has been shown that classical physics cannot give an adequate description of the theory of the atom and that the concept of quantized energy must be introduced.
    The idea that all particles have waves associated with them has been introduced, and it has been shown how this proposition has been demonstrated convincingly for electrons. Finally, there was the concept that the energy of an atom (and, incidentally, any physical system) is not a continuous variable but can only assume one of a set of discrete energy states or levels.

PROBLEMS

1.1 Assuming that the surface temperature of the sun is 6000 K, calculate the total energy radiated per second given that the sun's radius is 7.0×10^8 m. If the radius of the earth's orbit around the sun is 1.5×10^{11} m calculate the rate of arrival of energy per square metre on the earth's surface when the sun is directly overhead.

1.2 Derive Wien's law and the Stefan–Boltzmann law from Planck's radiation formula (1.4). (*Hint:* put $x = h\nu/kT$ and differentiate to locate the maximum value of I. Also note that the definite integral $\int_0^\infty [x^3/(e^x - 1)]\,dx = \pi^4/15$.)

1.3 A clean tungsten surface is irradiated with radiation of wavelength 187 nm and electrons of maximum energy 1.2 eV are emitted. Calculate the maximum wavelength that will cause photoelectric emission from the surface.

1.4 Photoelectric emission from caesium with light of wavelength 546.1 nm is just prevented by a retarding potential of 0.374 V, whilst for 312.6 nm the potential is 2.070 V. Calculate the value of the electronic charge.

1.5 Derive an expression for the de Broglie wavelength (in nanometres) of an electron in terms of the potential difference V (in volts) through which it has been accelerated.

1.6 (a) Find the wavelength of a 1 kg object whose speed is $1 \, \text{m s}^{-1}$.
(b) Calculate the de Broglie wavelength of yourself when you are walking at a brisk pace. Discuss!

1.7 (a) Show that for a free particle of velocity v and mass m the group velocity is the same as the particle velocity.
(b) By considering the phenomenon of beats show that the group velocity $v_g = d\omega/dk$. (*Hint:* add two sine or cosine waves which respectively have wavenumbers k and $k + dk$ and frequencies ω and $\omega + d\omega$, and assume that $dk \ll k$, and $d\omega \ll \omega$.)

1.8 An electron moves with a velocity of $500 \, \text{km s}^{-1}$, which is determined to an accuracy of 0.01%. With what fundamental accuracy can its position be located

1.9 The uncertainty principle can be expressed in the form $\Delta E \Delta t \approx h/2\pi$. If electrons remain in excited energy states for an average time of $10^{-8} \, \text{s}$, estimate the uncertainty in the energy of the excited level and the corresponding spread in frequencies if electrons undergo transitions from this level to the ground state energy level.

1.10 When you are doing experiments in the laboratory, is the uncertainty principle important or not?

1.11 The first four lines of the Lyman series have the following wavelengths:

    121.6, 102.6, 97.2 and 95 nm

Determine graphically the Rydberg constant for hydrogen.

1.12 What is the binding energy in electronvolts of an electron in the hydrogen atom for $n = 10$?

1.13 A hydrogen atom is de-excited from its initial state to the ground state, and in so doing emits a photon of wavelength 102.6 nm. What was its initial state?

1.14 Calculate the current corresponding to an electron moving in the first Bohr orbit.

2 Structure of Crystalline Solids

The previous chapter dealt essentially with phenomena associated with isolated atoms or collections of isolated atoms, for example in a gas emitting radiation. We will now turn our attention to questions relating to what happens when atoms are brought together to form a solid, and at the outset it is important for us to be clear what is meant by a solid in the context of this book, that is, to distinguish between crystalline and non-crystalline solids.

Essentially a crystalline solid is one in which the constituent atoms are arranged in an ordered array such as in a metal or a semiconductor, as opposed to a disordered array as in a glass, as shown in Figure 2.1. In the ideal crystalline solid, every atom is in its place and there is long-range ordering; in glass (which is essentially a frozen liquid) there is some short-range ordering, but no long-range order.

Although in recent times a great deal of interest has been shown in glassy materials, particularly in amorphous semiconductors, it is crystalline solids that are of primary interest to us. In fact, surprisingly few materials are amorphous,

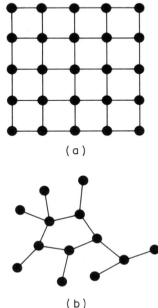

(a)

(b)

Figure 2.1. (a) An ordered array. (b) A disordered array

two common examples being glass and plastic. One distinguishing feature is the lack of a distinct melting point. As they are heated, they gradually soften, and the transition from solid to liquid is not readily discernible.

Several questions now arise: How are the atoms arranged in a given crystalline solid? What is the nature of the forces holding the atoms together? How do real solids compare with our 'ideal' models of solids? What experimental evidence is there to support any models that are developed?

Let us suppose initially that the atoms are like hard sticky spheres which, when placed in contact, stick firmly together without too much deformation. If, on the whole, the atoms remain in the positions in which they are initially placed, then the result is a solid, as opposed to a liquid or a gas, where the atoms are able to move freely through the whole region available to them.

2.1 STRUCTURE OF SOLIDS

If a large number of atoms are placed close together, then fairly obviously they can be arranged in an infinite number of disordered arrays. In fact, they might appear like a jar full of sticky spheres having no ordered arrangement but simply stuck together with gaps and voids between them. However, the minimum energy state corresponding to equilibrium requires that the atoms are in an ordered array in the solid state. In other words, to comply with physical law, the spheres should be arranged neatly in rows, the rows put together to form planes, and then the planes stacked to form an ordered three-dimensional shape as shown in Figure 2.2.

There is evidence in nature that atoms are indeed neatly arranged in this way. Many solids exist that are obviously crystalline (that is they have a regular appearance with definite crystal faces, corresponding pairs of faces having the same angles between them) such as rock salt and calcite (Figure 2.49). In fact, many solids which appear non-crystalline to the naked eye are seen to be definitely crystalline when viewed through an optical microscope. These solids consist of masses of small crystals randomly orientated with respect to each other and are termed polycrystalline.

It would be reasonable to assume that the regular outward appearance of crystals is but a manifestation of a regular internal arrangement of the atoms. This assumption has been demonstrated to be true by the methods of X-ray diffraction.

    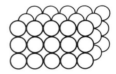

Figure 2.2. Diagram showing the ordered arrangement of atoms in solids

2.2   X-RAY DIFFRACTION BY CRYSTALS

If a beam of X-rays is allowed to fall onto a crystal and the scattered radiation from the crystal is recorded on a photographic plate, a well-defined pattern of spots is seen on the developed plate. This experiment was first carried out by W. Friedrich and F. Knipping (1912) at the suggestion of M. von Laue, who had noted that X-rays were thought to have a wave nature with a wavelength of about 10^{-10} m and that the spacing of atoms in solids was thought to be of the same order of magnitude.

The simple theory of diffraction by an optical grating (manufactured by ruling closely spaced lines on a glass sheet) indicates that the separation of the rulings should be of the order of magnitude of the wavelength to be analysed. If the line separation is too great, the maxima in the diffraction pattern are very close together and the dispersion is small. Hence, for the analysis of X-rays, a grating with spacings of about 10^{-10} m should be produced. This is technically impossible, and von Laue had the notion that if the atoms in crystals were, in fact, arranged in regular rows and planes, then a crystal might serve as a three-dimensional diffraction grating. The experimental arrangement used by Friedrich and Knipping is shown in Figure 2.3. X-rays from the X-ray tube were collimated into a narrow pencil by passing them through small holes in two lead screens. The pencil of X-rays fell on to a crystal of zinc sulphide. As anticipated, the photographic plate placed some distance behind the crystal exhibited a striking and complex diffraction pattern.

Thus, a completely new means of research into the structure of solids was originated. With X-rays of known wavelengths, crystal structures can be investigated in great detail: not only can the atomic arrangement be determined but also the electron distribution of the individual atoms.

W. L. Bragg (1913) gave a very simple interpretation of the spot arrangement on the photographic plate in terms of the wavelength of the X-rays used, the

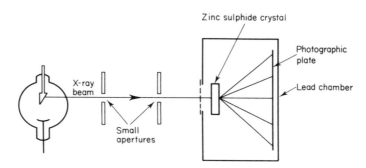

Figure 2.3. Experimental arrangement used by Friedrich and Knipping in investigating the diffraction of X-rays by crystals

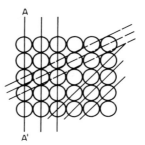

Figure 2.4. Planes of atoms in a crystal

separation of the planes of atoms in the crystal and the angle of incidence of the X-rays on the planes. If the crystal contains regularly arranged atoms, there will be many different sets of parallel planes (containing a large number of atoms) throughout the crystal. This is shown for a two-dimensional crystal in Figure 2.4. Each atom scatters the X-rays more or less uniformly in all directions but, because of the regular atomic arrangement, the scattered radiation from all atoms is in phase in certain directions and interferes constructively. In all other directions, there is destructive interference.

If the X-ray beam is incident at an angle θ to the planes, as shown in Figure 2.5(a), and then follows the direction BC or EF, also inclined at θ to the planes, the path lengths (AB + BC) and (DE + EF) are equal; thus the radiation scattered in this direction from atoms E and B will be in phase, and similarly for all atoms in the top plane. The radiation scattered from atom G in the second plane will also be in phase with that from the top plane if the path difference is a whole number of wavelengths.

The path difference is (XG + GY), which, as can be seen in Figure 2.5(b), is $2d \sin \theta$, where d is the separation of the planes. Hence the condition for constructive interference is

$$n\lambda = 2d \sin \theta \qquad (2.1)$$

where $n = 1, 2, 3, \ldots$. This relationship is known as the Bragg law. It indicates that diffraction intensities can only build up at certain values of θ, corresponding to a specific value of λ and d. This is so because the wavelets scattered from

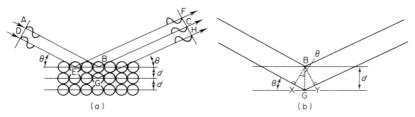

Figure 2.5. X-ray diffraction from a set of parallel planes showing (a) constructive interference of the scattered rays and (b) the construction for the Bragg law

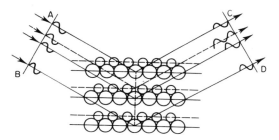

Figure 2.6. Diagram of a hypothetical crystal lattice showing how the atomic arrangement influences the observed diffracted beams

various points in the crystal have a common wavefront FCH only at these angles. As can be seen, the diffraction of X-rays by the atoms of a crystal can be treated as the reflection of X-rays by families of parallel planes in the crystal. By measuring the angles θ, the separation of these planes can be determined.

To see what effect the atomic arrangement has on the diffracted beams, let us consider Figure 2.6, which shows a hypothetical crystal composed of different sized atoms in two mutually displaced arrays. At an angle satisfying the Bragg law for both arrays, the atoms in each array scatter radiation in phase with other atoms of the same array. The radiations scattered by the two arrays, however, are out of phase because of the different path lengths between the wavefronts AB and CD. This has the effect of reducing the intensity of the diffracted beam in the direction θ in comparison with what it would have been had the crystal been composed of the large atoms only. From considerations such as this, the atomic arrangement in crystals can be determined from their X-ray diffraction patterns.

Example 2.1

In a Bragg spectrometer, X-rays of wavelength 0.154 nm are incident on a set of crystal planes in a sodium chloride (NaCl) crystal and the first-order diffraction maximum is observed at a glancing angle of 15.8°. What is the separation of the planes?

Solution. From equation (2.1)

$$n\lambda = 2d \sin \theta$$

For first-order diffraction $n = 1$; $\theta = 15.8°$ and $\lambda = 0.154 \times 10^{-9}$ m. Substituting:

$$0.154 \times 10^{-9} = 2 \times d \times \sin 15.8°$$

which gives

$$d = 0.283 \text{ nm}$$

X-ray diffraction, as indicated above, is analogous to reflection from a series of semi-transparent mirrors; it can also be considered as diffraction from a three-dimensional grating (similar to the diffraction of light from a one-dimensional grating). This approach leads to a series of three equations (the Laue equations) which have to be simultaneously satisfied for an intense beam of X-ray radiation to build up. This condition acts as a severe limitation on the number of such beams.

2.2.1   X-RAY DIFFRACTION—THE POWDER METHOD

To ensure that the Bragg law can be satisfied it is necessary to provide a range of values of θ or λ for a given set of d values. The various ways of doing this and the nature of the material to be investigated, single large crystal or polycrystalline, form the bases of the standard methods of crystal analysis using X-ray diffraction (for example the Laue, rotating crystal and Weissenberg methods). As an example of these, the powder method, which is widely used and easy to understand, will be considered.

The material to be investigated, in polycrystalline form, is placed in a narrow beam of monochromatic X-radiation. Among the large number of crystallites present, which have random orientations, certain crystallites must have the correct orientations to satisfy the Bragg law (here θ is the variable and the sample is often rotated to increase the range of θ values). The sample is placed at the centre of the powder camera and a photographic recording film in the form of a short cylinder coaxial with the sample axis records the diffracted beams as shown in Figure 2.7.

The diffracted X-rays leave the sample along lines lying on the surfaces of a set of cones whose common axis lies along the direction of the incident beam, each cone having a semi-apex angle 2θ, as shown in Figure 2.8. The cones intersect the film in a series of short arcs which are part of a concentric ring system (the Debye–Scherrer rings). Thus when the film is laid out flat it has the appearance of Figure 2.9. The arcs are symmetrically displaced on either side of the central beam direction and a measurement of their separation S enables the corresponding Bragg angle θ to be measured as $S = 4R\theta$, where R is the camera radius.

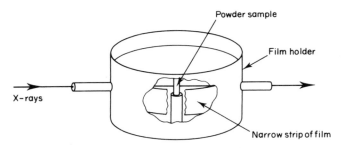

Figure 2.7. The powder camera

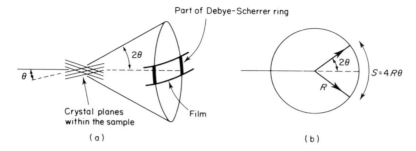

Figure 2.8. Diagrams showing: (a) how crystal planes inclined at θ to the incident beam give rise to reflections lying on the surface of a cone of apex angle 4θ; (b) the relationship between the diameter of the Debye–Scherrer rings and the radius of the camera

The separation d of the parallel planes which give rise to reflections at each value of θ can therefore be determined. From these measurements the lattice parameters can be obtained from which the size and shape of the unit cell (see below) of the crystal under investigation can be evaluated.

2.3 THE GEOMETRY OF PERFECT CRYSTALS

Measurements obtained from X-ray diffraction patterns of solids have enabled the atomic arrangement of crystalline substances to be determined. They have verified that atoms are arranged in regular arrays, but it is interesting to note that it is also possible to build up a great deal of information about the internal atomic arrangements in crystals by considering their symmetry properties. By observing their external appearance, the ease with which they can be cleaved along specific planes, and by measuring angles between their faces, for example, it is possible to deduce the *symmetry elements* of a given crystal and to classify it accordingly.

The essential problem is to find ways in which parts of an ideal crystal can be interchanged such that the result looks just the same as the original crystal. For any given crystal, there may be several ways—called *symmetry operations*—of doing this, and these will specify the *symmetry class* of that crystal.

Although crystals are clearly three dimensional, some of the salient features of symmetry operations can be illustrated quite adequately in two dimensions.

2.3.1 CENTRE OF SYMMETRY

The simplest type of symmetry element is the *centre of symmetry*. A crystal is said to have a centre of symmetry if to every point on one side of the crystal

Figure 2.9. A typical X-ray powder photograph showing the Debye–Scherrer rings produced by sodium chloride crystals

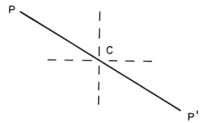

Figure 2.10. Centre of symmetry

there is a corresponding point on the opposite side which is equidistant from the centre and on the same line through the centre. This is shown in Figure 2.10, where the operation of the centre of symmetry at C produces the point P′ from P, with P′CP a straight line and P′C = PC. This can also be thought of as an *inversion*, through the centre C.

2.3.2 AXES OF SYMMETRY

Figures 2.11 (a), (b) and (c) illustrate twofold, threefold and fourfold symmetry axes, from which we can see that the degree of symmetry depends on how many times in one complete revolution about the axis the shape of the figure is restored. If that happens *n* times in one complete revolution about the axis O, the shape is said to have an *n*-fold symmetry axis. It turns out that in order to be consistent with a crystal's other symmetry requirements, only one-, two-, three-, four- and sixfold axes are allowed.

2.3.3 SYMMETRY PLANES

Unlike rotation about an axis, an operation that we can actually perform, there are other operations that we cannot. Such a *non-performable* operation is illustrated by reflection in a plane, as shown in Figure 2.12, in which point P′ is produced from point P by a mirror-like reflection in the plane. However, the problem of such operations can be overcome, since we can readily show

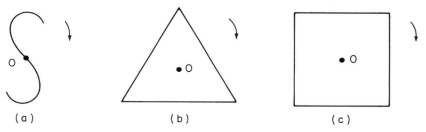

Figure 2.11. (a) Twofold (b) threefold and (c) fourfold symmetry axes

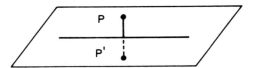

Figure 2.12. Plane of symmetry

that a succession of two non-performable operations is equivalent to a single performable operation. This is shown in Figure 2.13, in which the double reflection of the letter E is equivalent to a rotation about a single axis O.

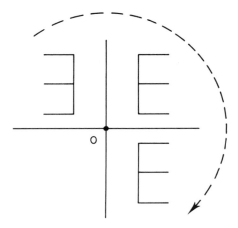

Figure 2.13. Equivalence of two non-performable operations to one performable operation

2.3.4   ROTATIONAL INVERSION SYMMETRY

Essentially this consists of a combined rotation and inversion to bring a crystal into self-coincidence. In Figure 2.14, the twofold rotation–inversion axis rotates P through 180° to intermediate position P′, and then inverts it through the centre C to P″.

These symmetry elements — axes, planes, centres and inversions — are called the macroscopic symmetry elements and occur singly or in groups in crystals and define the crystal external symmetry, the symmetry of its faces and its physical properties.

2.4   BRAVAIS LATTICES

By the middle of the nineteenth century crystallographers and mathematicians had collected a great deal of data relating to crystals and their macroscopic

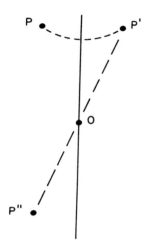

Figure 2.14. Rotation–inversion symmetry

symmetry elements. This work culminated in 1848, when A. Bravais demonstrated that there were only fourteen unique crystal (Bravais) lattices and, moreover, that, according to symmetry, all crystal structures could be placed in one of seven crystal systems or classes.

Imagine a lattice of lines dividing a three-dimensional space into equal-sized parallelepipeds which stand side by side with faces in contact so as to fill the space completely. The intersection of the lines are points of a space lattice. These are important points for frequently, but not always, they are the positions occupied by atoms in crystals or about which a group of several atoms is clustered. Since parallelepipeds of many different shapes can be drawn through the points of a space lattice to partition the crystal into small volumes called *unit cells*, the manner in which the network of lines is drawn is quite arbitrary (as can be seen for the 'crystal' shown in Figure 2.15).

The lines need not be drawn so that lattice points lie only at the corners of the unit cells; it is often more convenient to describe some crystals with respect

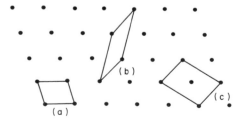

Figure 2.15. Possible unit cells formed by linking the points of a space lattice: (a) and (b) are primitive unit cells; (c) may be a more convenient way of describing a given structure and is called a conventional unit cell

to a unit cell in which points lie not only at cell corners but also at cell centres or at centres of the cell face, as shown in Figure 2.15(c). For the simplest cells, there is a one-to-one correspondence between lattice points and cells. Thus in Figure 2.15(a) and (b) each cell touches four points, and each point is shared between four similar cells, so each cell contains $4 \times \frac{1}{4} = 1$ lattice point; these cells are called primitive cells.

The important characteristic of a *space lattice* is that every point of the lattice has identical surroundings. The grouping of lattice points about any given point is identical with the grouping about any other lattice point in the lattice. Geometrically there are only fourteen unique ways in which this can be done — this gives the fourteen Bravais lattices, which are shown in Figure 2.16. It should be noted that the terms crystal lattice and crystal structure, while originally

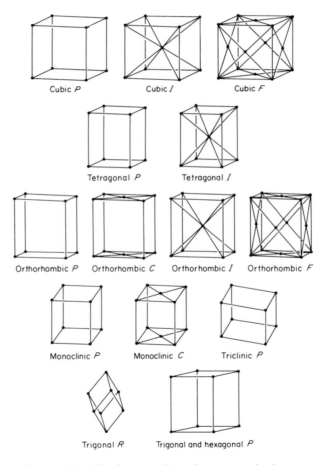

Figure 2.16. The fourteen Bravais or space lattices

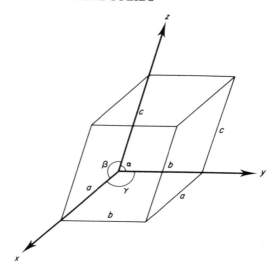

Figure 2.17. The crystallographic axes and lattice parameters

having different meanings, have been used wrongly as synonyms. There are only fourteen unique lattices, but there are a great many crystal structures, consisting of some fundamental pattern repeated at each point of a space lattice.

To specify a given arrangement of points in a space lattice or of atoms in a crystal structure, it is customary to give their coordinates with reference to a set of coordinate axes chosen with its origin at a lattice point. These 'crystallographic axes' are chosen for convenience for each crystal structure. For a cubic crystal, therefore, three mutually perpendicular axes of equal length, which form three edges of a cube, are used. Each space lattice has some convenient set of axes, and in fact seven different systems of axes are sufficient; these are the bases of the seven crystal systems. Referring to Figure 2.17, the unit lengths along the three axes are a, b and c, and the opposite angles are α, β and γ respectively. The parallelepiped formed as shown in Figure 2.17 often defines the conventional unit cell of a crystal. The crystal can be built up by stacking the unit cells, which are all identical in size and shape, in perfect alignment in three dimensions. When a crystal structure is determined, the values of the axial lengths and angles (the lattice parameters) are measured (that is the size and shape of the unit cell is determined).

2.5 MILLER INDICES

It is necessary to have a system of notation for the faces of a crystal, and for the planes within a crystal or space lattice, that will specify orientation without

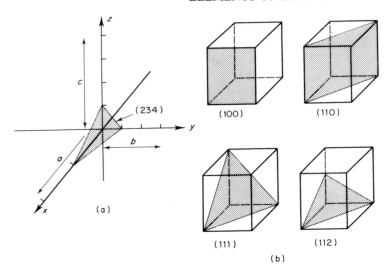

Figure 2.18. Miller indices: (a) an arbitrary plane; (b) some important planes

giving position in space. In 1839, W. H. Miller published a system that is now universally used for this purpose. These 'Miller' indices are based on the intercepts of a plane with the crystallographic axes. The intercepts are measured in terms of the dimensions of the unit cell, which are the unit distances along the three axes, and not in any specific length unit. Let us consider the plane shown in Figure 2.18(a) which cuts the x, y and z axes respectively at $\frac{1}{2}a$, $\frac{1}{3}b$ and $\frac{1}{4}c$ relative to the origin, regardless of the size of a, b and c. To determine the Miller indices, we adopt the following procedure:

(a) The intercepts on the three axes are found in multiples or fractions of the unit distances on each axis.
(b) In order to avoid fractions, the reciprocals of these numbers are taken (for important planes, these numbers are usually less than unity).
(c) The reciprocals are reduced to the three smallest integers having the same ratio.
(d) The resulting integers are enclosed in parentheses (hkl) to give the Miller indices of that and all parallel planes.

For the plane shown in Figure 2.18(a) the steps of the above procedure become:

(a) $\frac{1}{2}$, $\frac{1}{3}$, $\frac{1}{4}$.
(b) 2, 3, 4.
(c) 2, 3, 4.
(d) Miller indices (hkl) are (234).

Other examples are shown in Figure 2.18(b) where it can be seen that some of the shaded planes are parallel to one or two axes; that is the intercepts are infinity and thus the corresponding Miller index is zero (the reciprocal of infinity).

Parentheses (*hkl*) around Miller indices indicate a plane or a set of parallel planes. Curly brackets {*hkl*} signify planes of a 'form'—that is those planes that are equivalent in the crystal. For example, the faces of a cubic crystal {100} are (100), (010), (001), ($\bar{1}$00), (0$\bar{1}$0), and (00$\bar{1}$). The bar above some of the indices here simply indicates that the plane cuts the axis on the negative side of the origin, so the corresponding index is negative.

The indices of direction in the crystal are derived differently. If we consider a point at the origin of the coordinates, then any other lattice point can be reached by means of motion parallel to the three axes. If we suppose the desired motion can be accomplished by moving u times distance a along the x axis, v times b along the y axis and w times c along the z-axis, then [uvw] are the indices of the required direction. For example, the x-axis has indices [100], the y axis [010], and the z-axis [001].

The Bragg diffraction law (equation 2.1) gives the directions in which maxima are found when X-rays are diffracted from given crystal planes. This can be written more precisely as

$$n\lambda = 2d_{hkl} \sin \theta_{hkl} \qquad (2.1a)$$

where (*hkl*) refers to the particular set of parallel planes of interest.

2.6 CRYSTAL UNIT CELL STRUCTURES

Some of the important unit cell shapes are shown in Figures 2.19, 2.21, and 2.24. The usual convention has been followed in that the unit cell corners are located at the centres of atoms, though this is by no means essential. The diagrams show an 'exploded view' of the structures, in which the atoms are separated from each other and are joined by lines which give an idea of the

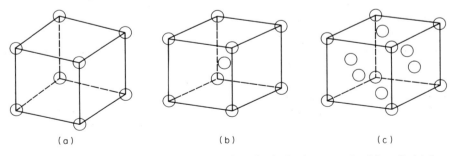

(a)   (b)   (c)

Figure 2.19. Cubic unit cells: (a) simple cubic cell; (b) body centred cubic cell: (c) face centred cubic cell

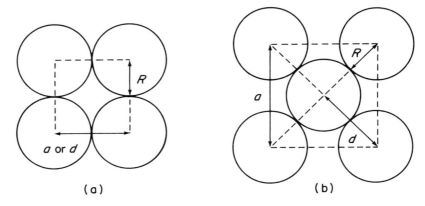

Figure 2.20. Cubic crystal face: (a) simple cubic; (b) face centred cubic

bonding forces (see page 51). A more accurate type of diagram is one in which atoms touch along their nearest neighbour distance, denoted by the letter 'd'. The lattice constant, which is the side length of the unit cell, is denoted by the letter 'a'. Half the nearest neighbour distance is defined as the atomic radius, R. This is illustrated for a (100) face of the simple and face centred cubic lattices in Figure 2.20. Although more realistic, these diagrams are a little difficult to interpret, for the symmetry is not always immediately obvious.

2.6.1 CUBIC STRUCTURES

Although only one element crystallizes in the simple cubic form (polonium) it is easy to visualize and is therefore often used as a basic example in discussing crystal structures. As can be seen from Figure 2.19(a), the unit cell contains one atom, since each corner is shared with seven other unit cells. The fraction of the volume occupied by atoms in a unit cell is called the packing density, which for the simple cubic unit cell is given by the volume occupied by one atom divided by the volume of the cell, which is readily shown to be $\pi/6$. The coordination number, that is the number of nearest neighbours, is six.

The reason that atoms do not tend to arrange themselves in a simple cubic structure is that it is a very open structure with many empty spaces. As a rule atoms tend to pack together more closely, which we can see, from Figures 2.19(b) and (c), is the situation in the body centred and face centred cubic structures, where some of the more obvious open spaces are occupied.

In the body centred cubic structure (for example Fe, K, Na, W) there are two atoms per unit cell ($8 \times \frac{1}{8} + 1$) and the coordination number is eight. For the face centred structure (e.g. Cu, Ag, Au, Al) the number of atoms per unit cell is four ($8 \times \frac{1}{8} + 6 \times \frac{1}{2}$) while the coordination number is twelve. Table 2.1 gives a summary of the more important parameters relating to the cubic systems.

Table 2.1. Summary of the more important parameters

Parameter	Simple cubic	Body centred cubic	Face centred cubic
Coordination number	6	8	12
Atomic radius	$a/2$	$a\sqrt{3}/4$	$a\sqrt{2}/4$
Packing density	$\pi/6$	$\pi\sqrt{3}/8$	$\pi\sqrt{2}/6$
Atoms per cell	1	2	4

2.6.2 HEXAGONAL STRUCTURE

We can envisage the hexagonal structure as being built up from layers of balls packed in triangular arrays; two such layers, one placed directly above the other, produce the hexagonal structure that has three atoms per unit cell and a coordination number of six. Figure 2.21 shows the relationship of the unit cell to the hexagonal Bravais lattice.

The most important variation of this lattice is the hexagonal close packed (h.c.p.) structure (for example Cd, Mg, Ti, Zn). The atom positions of the h.c.p. structure do not themselves form a Bravais space lattice; rather the structure can be thought of as a simple hexagonal space lattice with two atoms at each lattice point, one located at a reference origin 000 and the other at ($\frac{2}{3}a$ $\frac{1}{3}a$ $\frac{1}{2}c$) (this is equivalent to two interpenetrating simple hexagonal lattices). The h.c.p. unit cell, shown in Figure 2.22, contains six atoms and the coordination number is twelve, with each atom next to six atoms in its own layer and three in the layers above and below. This structure, together with the face centred cubic, corresponds to the most dense packing of atoms in space. Both structures can be visualized by packing equal spheres together in close packed layers in which each atom is in contact with six others. Such layers can then be stacked in

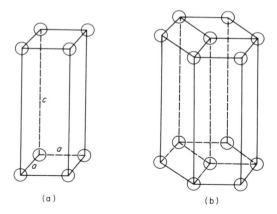

(a)   (b)

Figure 2.21. The hexagonal Bravais lattice (a) showing its relationship to a hexagonal unit cell (b)

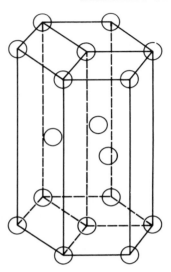

Figure 2.22. The hexagonal close packed structure

sequence. A second layer may be packed on the first so that each sphere in the second layer occupies a depression between three spheres in the first layer. There are, however, two ways of adding the third layer. The spheres in the third layer may be placed so that they are directly above those in the first, giving the ABABA . . . sequence of the h.c.p. structure, or they are placed above depressions in the first layer not occupied by second layer spheres. The spheres of the fourth layer then lie above those of the first in the ABCABCA . . . sequence of the face centred cubic structure (this layering sequence is seen in the layers of spheres normal to the direction of the body diagonal).

Example 2.2

Calculate the packing density for a hexagonal close packed lattice.

Solution. For an h.c.p. lattice, $c/a = \sqrt{8/3}$ and $R = a/2$.

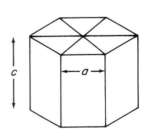
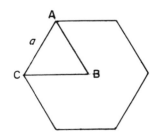

Volume of one atom $(4/3)\pi R^3 = (4/3)\pi(a/2)^3 = \pi a^3/6$
Volume of 6 atoms $= \pi a^3$
Volume of unit cell $= 6 \times$ (area of triangle ABC) $\times$ height

Volume of unit cell $= 6 \times \frac{1}{2}(\sqrt{3}/2)a \times c = \frac{(3\sqrt{3})}{2}a^2c$

Therefore the packing fraction $=$ volume of unit cell/volume of crystal

$$= \pi a^3 / \frac{1}{2}(3\sqrt{3})a^2c$$
$$= \frac{2}{3}\pi/\sqrt{3}\ a/c$$

But $c/a = \sqrt{8/3}$ so, packing density $= \dfrac{\sqrt{2}\pi}{6}$

Other crystal structures and their unit cells will be considered in relation to the bonding of the atoms discussed below.

2.7 ATOMIC BONDS

It has been shown in earlier sections of this chapter that the atoms in crystalline solids are arranged in neat ordered structures, but the nature of the forces which hold the atoms together (that is the nature of their 'stickiness') has not yet been described. The various ways atoms are bound together will now be discussed. Although the bonding of small groups of atoms can be classified under different types, this is not possible for solids because there are no solids in which the bonding is of one type only. However, as each type of atomic bond is described, consideration will also be given to a typical crystal in which the bonding is predominantly of that type.

Consider, first, the general situation of two identical atoms in their ground states being brought together from an infinite separation. The points of interest are the natures of the forces which come into play, whether these forces are attractive or repulsive and the energy of interaction of the atoms.

Initially, the energy of their interaction is zero. As the atoms approach, the attractive forces increase and the energy increases in a negative sense (the energy of attraction is negative since the atoms do the work, while that of repulsion is positive as work has to be done on the atoms to force them closer together). At a separation of a few atomic radii, repulsive forces begin to assert themselves, and the atoms reach an equilibrium separation r_0 at which the repulsive and attractive forces are equal and the mutual potential energy is a minimum. The situation is shown in Figure 2.23. The repulsive forces have a much shorter range than the attractive forces and are due partly to electrostatic repulsion of like charges and partly to the non-violation of the Pauli exclusion principle. As the atoms come closer together the Pauli principle would be violated unless some of the electrons move to higher energy states; the system therefore gains energy, a situation that is equivalent to the action of an interelectron repulsive force.

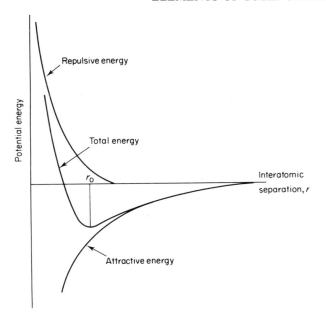

Figure 2.23. Graphs showing the interaction of two atoms as a function of their separation

In general, the total potential energy can be written as

$$E(r) = -E_{att} + E_{rep} \tag{2.2}$$

where E_{att} can be written as a small index power law:

$$E_{att} = \frac{-A}{r^n} \tag{2.3}$$

and E_{rep} as a high index power law or an exponential:

$$E_{rep} = B/r^m \quad \text{or} \quad B\exp\left(\frac{-r}{\rho}\right) \tag{2.4}$$

where ρ is a characteristic length that governs the rate at which the repulsive energy falls off with distance.

The values of the constants A and B, the indices n and m and the characteristic length ρ are governed by the actual nature of the bonding between the atoms, which is related to the charge distributions in the atoms making up the solid. However, in all cases the interaction provides the cohesive energy that brings the atoms of the solid together, which can be thought of as the difference between the atomic energy in the crystal and the energy of the free atoms.

Example 2.3

In a certain solid in which the equilibrium atomic spacing is 0.15 nm, the binding energy is 10% less than would be given for the attractive part alone. What is the characteristic length ρ in the equation

$$E(r) = -A/r^6 + B \exp(-r/\rho)$$

Solution. In equilibrium

$$dE/dr = 0 \text{ and } r = r_0$$

Therefore

$$\frac{6A}{r_0^7} = \frac{B}{\rho} \exp(-r_0/\rho) \tag{2.5}$$

and hence

$$E(r_0) = -A/r_0^6 + B \exp(-r_0/\rho)$$
$$= -A/r_0^6 \times 0.9$$

That is

$$B \exp(-r_0/\rho) = 0.1A/r_0^6 \tag{2.6}$$

Combining equations (2.5) and (2.6)

$$60\,\rho = r_0 = 0.15 \text{ nm}$$
$$\rho = 0.0025 \text{ nm}$$

2.7.1 IONIC BOND

The simplest interatomic force is the ionic bond, which results from the mutual electrostatic attraction of positive and negative charges. If an atom gains or loses one or more electrons it is said to be ionized and obviously it becomes either negatively or positively charged respectively. Atoms of many elements lose or gain electrons very easily because in so doing they acquire a completely filled outer electron shell which, as mentioned in Chapter 1, is a configuration of great stability. Thus sodium (Na) which has only one electron in its outer occupied shell (the 3s shell) will readily lose this electron to become a singly charged positive ion (Na^+). Similarly, calcium (Ca) will readily lose the two electrons in its outer occupied 4s shell to become the doubly charged calcium ion (Ca^{2+}).

Atoms such as those of chlorine, which have an almost completely full outer shell, readily accept additional electrons to fill the outer shell and thus become negative ions. Thus the building up of a sodium chloride molecule might be visualized as shown in Figure 2.24. However, it is incorrect to expect that Na^+ ions and Cl^- ions would link up in pairs as shown, because then there would be strong attractive forces within the paired ions of a sodium chloride crystal but negligible attraction between the pairs. As a result, solid sodium chloride would not exist.

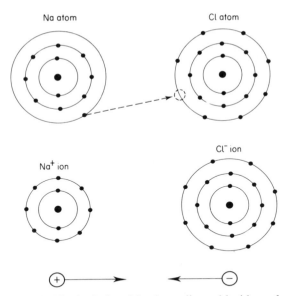

Figure 2.24. The ionic bond in the sodium chloride molecule

Actually, a negative charge possesses an attraction for all positive charges in the neighbourhood, and vice versa. Consequently, Na$^+$ ions will surround themselves with Cl$^-$ ions and Cl$^-$ ions with Na$^+$ ions in the crystal in such a way that the attraction between neighbouring unlike charges exceeds the repulsion due to like charges. The resulting sodium chloride structure is shown in Figure 2.25, which shows that there are two interpenetrating face centred cubic structures, one formed from sodium ions and the other from chlorine

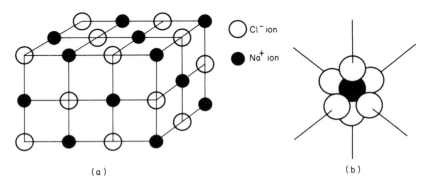

Figure 2.25. Formation of a typical ionic crystal (sodium chloride): (a) a face centred cubic structure of Na$^+$ ions interpenetrates a similar structure of Cl$^-$ ions; (b) a more realistic representation of Cl$^-$ ions grouped about a Na$^+$ ion

ions. In general, the ionic bond is quite strong, as evidenced by the fact that many ionic solids are hard and have quite high melting points (for example lithium fluoride and sodium chloride which have melting points of 845 and 801°C respectively).

The interaction energy for an NaCl molecule can be written as

$$E(r) = \frac{-e^2}{4\pi\epsilon_0 r} + B \exp\left(\frac{-r}{\rho}\right) \tag{2.7}$$

If the total number of molecules in the crystal is N and n the number of nearest neighbours of a given ion, the total interaction energy is

$$E_T = N\left[nB \exp\left(\frac{-r}{\rho}\right) - \frac{\alpha e^2}{4\pi\epsilon_0 r} \right] \tag{2.8}$$

where α is a geometrical factor called the Madelung constant. It can be calculated by summing the individual Coulombic attractions with successive shells of nearest neighbours, and this varies from structure to structure. For NaCl, α is 1.748.

Clearly the value of the equilibrium separation r_0 can be found by minimizing this expression. For NaCl, r_0 is 0.28 nm.

2.7.2   COVALENT BOND

Another way in which the outer electron shell of atoms can be effectively filled to achieve a stable configuration is by the sharing of electrons — this gives rise to the covalent bond. For example, chlorine requires one electron to complete the outer shell. This can be accomplished if two chlorine atoms join together and share their outer electrons. Figure 2.26 shows the resulting bond which enables the chlorine molecule to be formed.

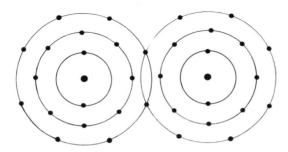

Figure 2.26. The chlorine molecule. By sharing electrons, both atoms effectively fill their outer electron shells

The atoms taking part in covalent bonding need not necessarily be alike. For example, hydrogen and chlorine may combine as follows:

$$H:H + Cl:Cl = 2 H:Cl$$

where the dots represent shared electrons in the outer orbit.

From the point of view of this book, the most significant crystal structure resulting from covalent bonding is the diamond structure, as the important semiconductors silicon and germanium crystallize in this structure (this results from the similarity of their outer electron shells). The carbon atom has four electrons in its outer shell ($n = 2$) which can accommodate a total of eight electrons. In the diamond structure each carbon atom can be considered to be at the centre of a regular tetrahedron with its nearest neighbours at the corners of the tetrahedron, as shown in Figure 2.27(a). Each of the four corner atoms shares one of its outer electrons with the central atom, thereby effectively making up a closed shell of eight electrons. The central atom, of course, shares each of its four outer electrons with one of the corner atoms, thereby contributing to their closed shells. Each corner atom in turn can be regarded as being at the centre of a tetrahedron so that the structure consists of a system of linked tetrahedra as shown in Figure 2.27(c).

The structure is in fact face centred cubic (FCC) with a basis of two atoms at each lattice point, one on the lattice point 0,0,0, the other at $a/4$, $a/4$, $a/4$. An equivalent representation is of two interpenetrating FCC lattices which are offset with respect to each other by a quarter of the body diagonal.

The covalent bond is usually very strong so that covalent crystals are usually hard and have a high melting point. Indeed, diamond is the hardest material and it has the highest known melting point at more than 3550°C.

If the two interpenetrating FCC lattices mentioned above are comprised of different elements the atoms on the two sub-lattices are of course no longer equivalent but they still form a tetrahedral structure. This is called the zinc blende structure (ZnS) and it is typical of such important semiconductors as gallium arsenide (GaAs), gallium phosphide (GaP) and indium antimonide (InSb).

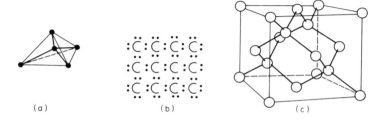

(a)   (b)   (c)

Figure 2.27. The structure of diamond: (a) the tetrahedral bonds; (b) schematic representation showing the bonding electrons; (c) the crystal structure

2.6.3 METALLIC BOND

In contrast to ionic and covalent bonds, the properties of the metallic bond cannot be inferred from the nature of bonding in isolated molecules and a model or bonding in metals is not easy to construct. In some respects, however, it may be regarded as intermediate in character between that of ionic and covalent bonds.

The metal atoms readily lose their valence electrons (thereby acquiring the stable closed shell electron configuration of the inert gases) to form positive ions. The valence electrons may then be regarded as being shared by all of the ions in the crystal. In other words the freed valence electrons form an electron 'gas' which may be regarded as permeating the whole crystal, as shown in Figure 2.28. The bonding is then essentially an electrostatic interaction between the array of positive ion cores and the electron gas. The free electrons of this model explain the high electrical and thermal conductivities of metals together with their high ductility (see Section 3.2).

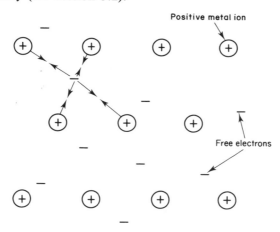

Figure 2.28. Schematic representation of metallic bonding. There is an attraction between each free electron and neighbouring metal ions

2.7.4 OTHER BONDS

There are several secondary bonding mechanisms which give rise to rather weak interatomic forces. Of these, the van der Waals bond is probably the most important and is almost entirely responsible for binding the atoms together in solidified inert gases. Van der Waals forces are essentially electrostatic in nature, and they arise as a result of momentary shifts in the relative positions of the nucleus and electrons of an atom. As a result, the atom forms an electric dipole which in turn induces dipoles in neighbouring atoms. The attraction between charges of opposite sign in the dipoles thus gives rise to the attractive forces.

2.8   REAL CRYSTALS

Several of the questions posed at the beginning of the chapter have now been answered, the orderly nature of the atomic arrangement in crystals having been emphasized throughout. We shall now answer another of the initial questions by considering how real crystals compare with the rather perfect models we have suggested, in which every atomic site is filled and there is no disturbance of the atomic planes.

Shortly after Laue and Bragg began their work on the determination of the atomic arrangement in crystals, it became apparent that variations of certain observed properties in solids (such as plasticity, crystal strength and electrical conductivity) could not be explained on the basis of differences in crystal structure alone. Several workers suggested that crystals are not perfect but that they contain defects. Several different types of defects have now been identified, and it has been realized that they have a marked effect on both mechanical and electrical properties. For example, calculation and measurement of the ideal shear strength of copper and aluminium differ by a factor of about one thousand. At first it was thought that the theoretical calculation was incorrect, but it is now recognized that it is the presence of dislocations in the crystal that greatly reduces the strength of pure copper and aluminium. In recent years, metal crystals have been grown in the form of thin 'whiskers' which have a high degree of perfection. These whiskers are sometimes almost free of dislocations, and their strength is near the theoretical value.

2.9   DEFECTS IN CRYSTALLINE MATERIALS

2.9.1   POINT DEFECTS

All the atoms in a perfect crystal are at specific atomic sites. In real crystals, however, given atomic sites may not be occupied, that is there may be *vacancies*. On the other hand, there may be atoms in *interstitial* positions. These point defects are illustrated in Figure 2.29 for a simple cubic lattice. The vacancy is simply formed by the removal of an atom from an atomic site, while the

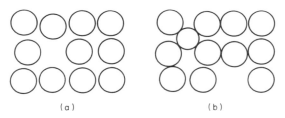

(a)   (b)

Figure 2.29. (a) Vacancy or Schottky defect; (b) interstitial or Frenkel defect. Note the deformation of the lattice caused by the defects

interstitial defect is formed by the introduction of an extra atom into a non-regular atomic site. The interstitial atom may be an atom that originally occupied a vacancy, in which case the defect is called a Frenkel defect. In ionic crystals like sodium chloride, the removal of an ion produces a local charge as well as a distortion of the lattice. To conserve overall charge neutrality, the vacancies occur either in pairs of oppositely charged ions or in association with interstitials of the same sign.

Impurity atoms in otherwise pure crystals can be considered as point defects, and they play an important rôle in the electronic and mechanical properties of all materials. Impurity atoms may either occupy a normal atomic site in the parent lattice (a *substitutional impurity*) or a non-regular atomic site (an *interstitial impurity*). These are shown in Figure 2.30. All point defects produce a local distortion in the otherwise perfect lattice, the amount of distortion depending on the crystal structure, parent atom size, impurity atom size and crystal bonding. These local distortions act as extra scattering centres to the flow of electrons through the crystal—thus increasing the resistance of the crystal.

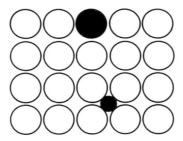

Figure 2.30. Substitutional and interstitial impurity atoms

Point defects have associated energy levels which lie between the bottom of the conduction band and the top of the valence band. Moreover, since a charge is usually associated with point defects, they can act as *traps* for current carriers which then remain bound at such traps and form further lattice defects. These may, as in 'colour centres', have their own set of energy levels such that electrons can be excited from the ground level of the trap to excited levels, giving rise to the absorption of radiation and possibly causing the solid to appear coloured. The positions between the bands of some of the common lattice defects in an alkali halide crystal are shown in Figure 2.31.

2.9.2 LINE DEFECTS

The concept of *line defects* or *dislocations* arose primarily from a study of plastic deformation processes in crystalline materials. There are two basic types of dislocation, edge and screw dislocations, which are illustrated in Figure 2.32.

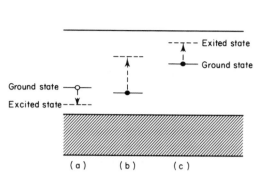

Figure 2.31. Energy bands in an alkali halide showing energy levels: (a) for trapped holes (V centres); (b) for impurity centres; (c) for trapped electrons (F centres)

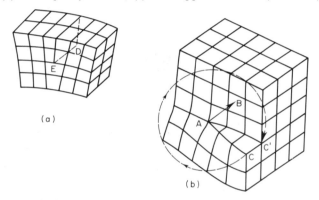

Figure 2.32. Crystal planes in the neighbourhood of: (a) an edge dislocation ED; (b) a screw dislocation AB

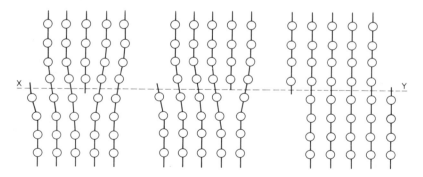

Figure 2.33. Diagram showing how the presence of an edge dislocation facilitates slip in a crystal (slip occurs along XY)

The edge dislocation can be visualized as an extra plane of atoms inserted part way into the crystal. The edge of this extra plane, ED in Figure 2.32(a), is the actual dislocation. There is severe distortion in the region around the dislocation and the lattice planes are bent. The presence of the edge dislocation greatly facilitates slip in the crystal when a shear force is applied, the slip occurring normal to the line marking the end of the extra plane of atoms ED. The movement of an edge dislocation is shown in Figure 2.33.

The screw dislocation occurs when one part of the crystal is displaced relative to the rest. In Figure 2.32(b), AB is the screw dislocation—so called because if the line AC is rotated about AB through 360° while being kept on an unbroken crystal plane, then C will advance one atomic spacing in the direction of the dislocation AB to C′, and each subsequent rotation will further advance the point C.

Owing to the distortion of the lattice caused by edge and screw dislocations, which usually occur together, there will be abnormally high stresses in their neighbourhood. If an external stress is applied, its effects will be pronounced at the dislocations because at these points the crystalline binding forces will be exceeded first, causing the crystal planes to slip. A rough analogy might be the tearing of a piece of paper. If the paper is pulled uniformly from opposite edges, it can withstand quite high stresses. However, if the stress is concentrated at one point, as in a tearing motion, it is easy to rupture the paper since the point of concentrated stress propagates along the edge of the tear.

It should be noted that dislocations can be blocked, just as the tear in the paper can be stopped by sticking a piece of tape across it. Motion along the dislocations can be prevented by the presence of dispersed impurity atoms or by other dislocations which may have been produced by cold work of the metal, for example by machining processes. Grain boundaries (see below) also play a part in these 'pinning' processes. These and other techniques are used in improving the mechanical strength of metals.

A simple device for studying point and line defects is the bubble raft, in which the surface of a liquid is covered with a layer of uniformly sized bubbles. By the application of forces to different regions of the raft, various defects can be produced. Examples of these are shown in Figure 2.34. However, it should be noted that dislocations can easily be revealed with etching techniques. Etching involves chemical treatment of a solid surface with a suitable solvent to produce a very high-quality surface finish. Examination of the surface by means of a high-power optical microscope or scanning electron microscope after etching enables the observer to see etch pits, which represent the points at which dislocations emerge on the crystal surface. Owing to the lattice distortion and resulting weakening of the atomic bonds, the etch solutions act more vigorously in the regions of dislocations. Etch pits in lithium fluoride are shown in Figure 2.35.

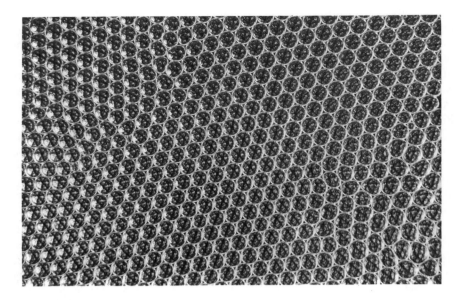

Figure 2.34. Edge dislocations as illustrated by a bubble raft (the dislocations can be seen most easily by viewing from the lower left-hand side almost parallel to the page). Crystal grain boundaries are also simulated with the bubble raft (upper left); they can be seen to be made up of a number of dislocations

2.9.3 PLANE DEFECTS

The predominant plane defect is the *grain boundary* (Figure 2.34), which occurs in most materials. Most crystalline solids do not consist of a large single crystal but of many randomly orientated crystallites. The join of these crystallites or grains results in grain boundaries, which mark mismatches in the rows and planes in the two adjoining crystallites. Each grain in itself is a single crystal and probably contains the point and line defects already described.

Grain boundaries block movement along dislocations by providing an effective edge to the crystal where, of course, movement must stop. It is well known that polycrystalline materials are invariably more resistant to fracture than single crystals of the same material.

Another plane defect is the *stacking fault*, which occurs when mistakes are made in the sequence of stacking the crystal planes. The plane separating two incorrectly juxtaposed layers is the stacking fault, an example of which is shown in Figure 2.36. Stacking faults occur most readily in crystals in which the layer sequence is ABCABC—for example in the face centred cubic structure. They are not found so often in structures such as the body centred cubic, in which the layer sequence is ABAB.

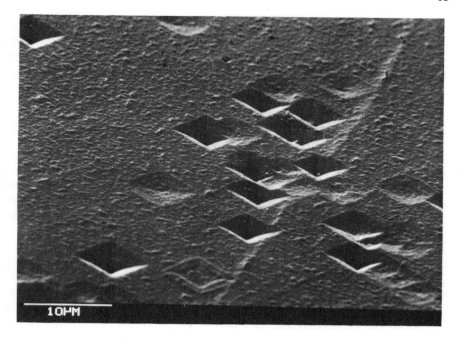

Figure 2.35. Etch pits in lithium fluoride

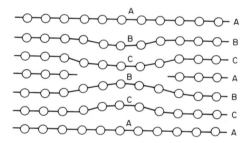

Figure 2.36. Stacking fault

2.9.4   PHONONS

It has already been seen (Section 1.1) that the atoms of a solid can be regarded as three-dimensional simple harmonic oscillators and that at all temperatures above absolute zero the atoms suffer a periodic displacement from their mean, equilibrium positions. The energy of such oscillators is quantized, and it can be shown that the allowed energies are given by

$$E_n = (n + \tfrac{1}{2})h\nu \qquad\qquad (2.9)$$

where $n = 0, 1, 2, \ldots$ and ν is the mechanical frequency of vibration. The energy change accompanying a transition from one energy state of quantum number n_1 to the next highest state described by quantum number n_2 is given by

$$\begin{aligned}
E_{n_2} - E_{n_1} &= (n_2 + \tfrac{1}{2})h\nu - (n_1 + \tfrac{1}{2})h\nu \\
&= (n_2 - n_1)h\nu \\
&= h\nu
\end{aligned}$$

as n_1 and n_2 are any two integers differing by unity. Hence, when thermal energy is absorbed by an atom, the energy of the atom increases by $h\nu$; if the transition occurs in the reverse order, energy is emitted. The situation is entirely analogous to the transitions involving the absorption and emission of photons, and the name phonon is used to describe the quantum of thermal energy absorbed or emitted by an atom.

The vibrations of the atoms which are coupled together can be envisaged as giving rise to vibrations of the whole crystal, that is to lattice vibrations. The energy of the whole vibrating system is still quantized and can only change by amounts $h\nu$, only now the quanta of vibrational energy can be regarded as increasing the vibrational displacements of all the atoms and not just one of them. Thus, if the temperature is raised the lattice will absorb phonons, thereby increasing the amplitude of the atomic vibrations. Hence phonons can be classified as crystal imperfections since they distort the lattice from its ideal state in which all the atoms are at rest. It can therefore be seen that no matter how carefully a crystal is prepared there will always be one form of defect present, namely phonons.

Phonons give rise to electron scattering and therefore increase the electrical resistivity of solids as the temperature rises. They are also responsible for part of the thermal conductivity of all solids and in insulators, such as diamond, they are almost entirely responsible for thermal conduction. There is a greater phonon density in a hot region of a crystal than in cooler regions and one thinks of heat conduction as being due to the diffusion of phonons down the temperature gradient from the hot to the cold regions. (In metals there is, of course, a substantial contribution to the conduction of heat by the free electrons.)

2.10 CRYSTAL GROWING TECHNIQUES

Many crystal properties are dependent on the ordered periodic arrangement of the atoms in a lattice. Very few materials occur naturally as large, pure single crystals, and thus crystals for research and use in solid state electronic devices have to be grown synthetically. Natural crystals contain many impurities, but

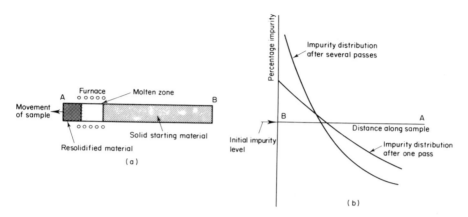

Figure 2.37. Zone refining: (a) sketch of the experimental arrangement; (b) redistribution of the impurity as a result of passing the molten zone from A to B

in the synthetic process one can start with very pure material (impurity content less than one part in 10^{13}) and add known impurities as desired to increase a specific property of the crystal such as electrical conductivity. In order to produce very pure starting material, a process known as zone refining is used on material that has been purified chemically to a purity of 99.999%.

Zone refining, shown schematically in Figure 2.37(a), is simply fractional solidification in which a solid forming from an impure liquid contains less impurity than the liquid. A long crucible is filled with the material in the form of a bar, and the crucible is then pulled slowly through a furnace in which only a short region is maintained at a temperature above the melting point of the material. Thus, as the crucible is moved slowly through the furnace, a molten zone moves along the bar. The zone melts impure material in front of it, while relatively pure material solidifies behind it; in effect, the molten zone sweeps the impurity along with it. The effect on the impurity level is shown in Figure 2.37(b), from which it is obvious that several passes of the zone through the material may be necessary to produce the required degree of purity. The impurity collects at end B of the rod, this end being cut off and repurified chemically.

The purified material can be used to prepare single crystals by one of many techniques. The choice of a technique for a particular material is a matter of trial and error, crystal growing being more of an art than a science. Three of the most used methods are described below. Other techniques include growth from solution, growth from the vapour phase (for example cadmium sulphide), growth under very high pressure (for example diamond) and growth under hydrothermal conditions (for example quartz).

In order that the properties of the grown crystal approach the ideal as nearly as possible, we endeavour to keep the defect concentration as low as possible by having a large single crystal which is free from grain boundaries, thermal strains, dislocations and point defects. To achieve this, the crystal is grown very slowly under carefully controlled thermal conditions.

We can understand the reason for these conditions if we think back to experiments in the school chemistry laboratory in which rather beautiful crystals of chemicals such as copper sulphate were produced. A small 'seed' of copper sulphate is suspended in a very concentrated solution and is left there for several days. In this time, the crystal grows by the addition of atoms to the external faces of the seed. The atoms are moving around in random directions in the solution, and it is only when the atom presents itself at an atomic site on the face in the correct way that it will 'stick' there. A similar situation exists in growth of crystals from the melt, in which atoms move around until they present themselves correctly at the growing face. The temperature of the growing solid is one or two degrees below the crystal melting point, while that of the liquid melt is a few degrees above; these conditions have to be maintained as uniformly as possible during growth. A full treatment of the growth mechanism of crystals is beyond the scope of this book, though an outline of some of the more important techniques is given below.

2.10.1   THE CZOCHRALSKI METHOD

The Czochralski or Kyropoulos method is commonly used for producing crystals of silicon and germanium. A seed crystal is dipped into a crucible of the molten material and then slowly withdrawn as new lattice layers solidify on the crystal surface. A typical arrangement is shown in Figure 2.38. The crucible is often a hole in a graphite block which can be heated with a high-frequency induction coil, or alternatively a crucible of another material, usually silica, may be placed inside the graphite. To help in keeping the temperature and crystal growth uniform the growing crystal and/or the crucible are rotated at a few revolutions per minute. The rate at which the crystal may be pulled from the melt depends very much on the material and the degree of perfection required. A few millimetres per hour is a typical rate. Using this technique it is now possible to grow silicon crystals up to a diameter of about 20 cm (Figure 2.49).

2.10.2   THE BRIDGMAN–STOCKBARGER METHOD

The Bridgman–Stockbarger method is shown in Figure 2.39. The crystal is grown in a crucible with a conical tip, which stands on a cooled support. After the material has melted in the upper furnace, the crucible is lowered into the second furnace, which is maintained at a few degrees below the melting point. As the crucible passes through the steep temperature gradient, the material solidifies in the point of the cone which acts as a seed: a single crystal may form and grow to fill the whole crucible. A modified crucible (Figure 2.39c) can be used

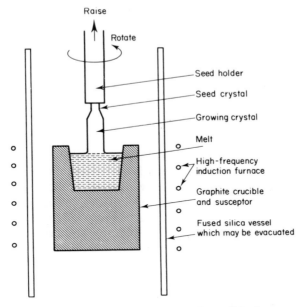

Figure 2.38. A Czochralski crystal growing furnace

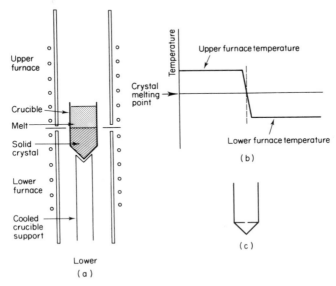

Figure 2.39. The Bridgeman–Stockbarger crystal grower: (a) experimental arrangement; (b) the temperature gradient along the axis of the furnaces; (c) the modified crucible

to select one of the many crystals which may form, as only one crystal is able to grow through the small hole in the baffle. This method is useful for growing crystals of dense materials, for example lead iodide, where the seed may be unable to support the growing crystal.

2.10.3 THE FLOATING-ZONE METHOD

Both of the above methods have the disadvantages that the crucible may contaminate the crystal. This can be avoided by using the floating-zone method shown in Figure 2.40. This method is a modification of the zone refining process. The starting material, prepared in the form of a polycrystalline rod, is held vertically in the furnace. A small region of the rod is melted and the molten zone can be moved through the rod from the seed to the other end by moving either the heater or the supports. The molten zone is held between the unmelted ends of the rod by surface tension. The floating zone is often used in zone refining to avoid the danger of contamination by the crucible. Zone refining and crystal growth by the methods outlined above are usually carried out in an inert atmosphere of argon or helium to prevent oxidation and additional contamination by impurities.

2.11 DEVICE FABRICATION

Since 1960 semiconducting devices have mostly been manufactured using a *planar epitaxial technique*. This method not only improved the manufacture

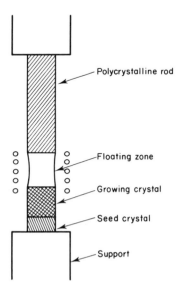

Figure 2.40. The floating-zone crystal growing method

of individual, discrete components such as junction diodes and transistors but also made possible the rapid development of modern integrated circuits, which may contain many components. Whether the aim is to fabricate a simple diode or a large and complex integrated curcuit, however, the basic processing requirements are essentially the same.

The basic processes of planar technology are epitaxial growth, oxidation, impurity diffusion, oxide etching and pattern definition using photolithography and metallization, though there are alternatives to some of these processes. We shall now discuss the processing steps in relation to the fabrication of silicon devices, but it should be noted that devices can be prepared by similar techniques in other semiconductors such as germanium and gallium arsenide.

The starting point is a flat polished slice of silicon which may be up to 150 or 200 mm in diameter. These slices or *wafers* are cut from single crystal ingots of silicon grown by the Czochralski method outlined above. The silicon is melted in a quartz crucible at 1425°C. A seed crystal usually orientated so that the (111) planes are normal to the axis of the crystal grower is dipped into the melt and a single crystal is grown at 20 to 150 mm per hour. The (111) orientation produces crystals containing fewer defects than other orientations, and also enables more devices per unit area to be formed. Wafers cut from a crystal with (100) orientation, however, are preferred for MOST devices (see Chapter 6). Wafers with this orientation give rise to a lower threshold voltage.

Carefully selected amounts of impurities are added to the melt to provide a crystal of the required type and electrical properties. The crystal ingot is cut into wafers using a diamond saw; the wafers are 0.25 to 0.5 mm thick depending on their diameter and the technology to be applied. The wafers are mechanically and chemically polished to remove any damage produced in the cutting operation to give a surface that is flat to within 5 μm; the wafer is at this stage often referred to as a substrate.

A thin *epitaxial layer* is now grown on the substrate by, for example, passing a mixture of hydrogen and silicon tetrachloride over the silicon wafers held at a temperature of about 1200°C. The silicon tetrachloride is reduced by the hydrogen on the surface of the substrate to release solid silicon which adopts the single crystal structure of the substrate. (Epitaxial comes from the Greek: *epi* = upon and *taxos* = arranged.) The epitaxial (or epi-) layer grows at about 1 μm per minute and layers of 1 to 20 μm are typically grown, though for some high-voltage devices the layer may be up to 120 μm thick. The epi-layer may be doped as it is produced by admitting appropriate amounts of phosphine (PH_3) or diborane (B_2H_6) gases into the reaction chamber to produce n-type or p-type material respectively. A typical impurity distribution resulting from this process is illustrated in Figure 2.41 which shows the formation of a semiconductor pn junction. As will be seen in Chapter 6, it is pn junctions such as this that provide the basic dynamic regions of the majority of semiconducting devices.

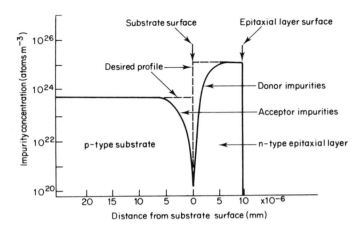

Figure 2.41. The impurity distribution resulting from the formation of a n-type epitaxial layer on a p-type substrate

It is important to realize that the entire device and indeed entire monolithic integrated circuits eventually are fabricated within the epitaxial layer. *Monolithic* (Greek: single stone) circuits are so called because all the active and passive components that it may contain are formed as an integral part of a single semiconducting substrate—here silicon.

The next step is to oxidize the surface of the epi-layer by heating the wafer to about 1100°C in steam or oxygen. The oxide layer performs three functions. Firstly, openings can be etched into it by photolithographic techniques and selected impurities diffused into the exposed areas to change the electrical properties of those areas while leaving the silicon under the oxide unchanged. Secondly, the oxide 'passivates' pn junctions that may exist at the surfaces of planar devices by protecting those active regions from contamination. This can give rise to so-called *surface states* which are discussed in Section 6.9.2. Thirdly, the oxide acts as a high-quality insulator, allowing interconnection conductors to pass over areas that would otherwise be electrically short circuited.

The *photolithographic process* is the technique used to define intricate patterns in the oxide layers and interconnection metal films. The surface of the wafer is coated with a photosensitive material called photoresist, which changes its structure on exposure to ultraviolet light. A photographic mask containing dark areas where an opening is to be formed is placed over the resist and the wafer is illuminated with ultraviolet light. The resist is then developed and the unexposed areas washed off. The wafer is then immersed in a strong etch such as hydrofluoric acid to dissolve the oxide and create an opening to the underlying silicon. The excess resist is then removed leaving the substrate ready for the diffusion stage.

This procedure of photoresist–mask–exposure–etch is often referred to as 'window opening' and is fundamental to device and integrated circuit technology. It may be repeated many times in the processing of an individual device.

The masks are made by photographic reduction of a master drawing of the windows down to the actual size of the windows to be opened. As the component density in complex integrated circuits increases, the tendency is for the masks to be made by electron beam machining, thereby eliminating the photographic reduction steps. The small wavelength of the electrons leads to very high resolution and electron lithography can produce masks with detail of around 0.1 μm wide.

When a window has been opened in the oxide in the appropriate place, impurity atoms can be diffused into the semiconductor through the window (Figure 2.42). The diffusion is normally carried out at temperatures of 850 to 1200°C. The wafers are held in quartz jigs and the dopants are introduced into the furnace with carrier gases. In most cases an extremely thin layer of dopant-rich silicon is created on the surface of the wafer and then redistributed by diffusion to the required depth. Diffusion is time, temperature and concentration dependent and it also varies for different impurities; hence an almost infinite number of redistribution profiles can be obtained. By manipulating these profiles one arrives at the optimum electrical properties for a particular device. The aim of diffusions of this type is often to produce a pn junction in the epi-layer. This is achieved by *compensation doping* in which an original n-type region (say) is converted to p-type by diffusing in an excess of acceptor impurities. Junction formation by compensation doping is illustrated in Figure 2.43. As with all stages in the fabrication of devices careful temperature control and extreme cleanliness is vital for reproducible production.

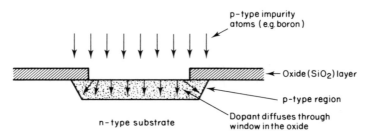

Figure 2.42. Oxide masking and diffusion in planar geometry

Alternative methods of introducing dopants into the surface layer of the silicon wafer have been developed. Of these *ion implantation* has proved to be the most useful. It has the advantages of being performed at room temperature and of producing a very high degree of lateral uniformity in the doped region. In this technique atoms of the required dopant are ionized, accelerated to high velocities

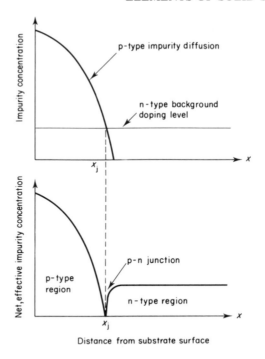

Figure 2.43. The p–n junction formation by compensation doping

by an electric field in a vacuum chamber and directed to the appropriate regions of the wafer. The ions penetrate the surface and form a shallow layer that can either be used directly or thermally redistributed.

Following implantation, the wafer is annealed to reduce crystal damage and to ensure that the implanted ions are able to move into the crystal sites normally occupied by the parent silicon atoms, that is form substitutional impurities. If the implanted ions are left in the interstitial positions which they tend to occupy on implantation then they are not electrically active in terms of producing additional carriers. Annealing takes place at about 800°C; at this temperature the rate of diffusion is very low so that there is very little redistribution of the implanted impurity profile.

When the doping is completed the surface is reoxidized and the whole process repeated as required. Thin films (1–2 μm) of metal, usually aluminium, are defined into patterns of photolithography to form interconnecting tracks and connections to the final device. The thin metal films are usually formed by thermal evaporation in a vacuum.

It is important to stress that discrete devices or even integrated circuits are not prepared individually. A 150 mm diameter wafer may, for example, have about 25 000 integrated circuits fabricated on it simultaneously, each integrated circuit itself comprising several diodes, transistors and passive components. Similarly, in some of the processing steps described above, for example in forming the epi-layer, up to 30 wafers may be treated simultaneously. Figure 2.44 illustrates this situation.

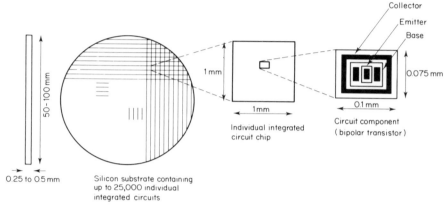

Figure 2.44. Typical geometry of silicon wafer. Circuit chip and component

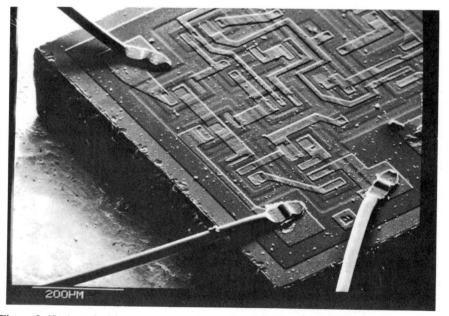

Figure 2.45. A typical integrated circuit showing the silicon chip and electrical leads-in (the area of chip occupied by a junction transistor is typically about 200 μm $\times$ 150 μm — such an area can be seen immediately to the left of the lower lead-in)

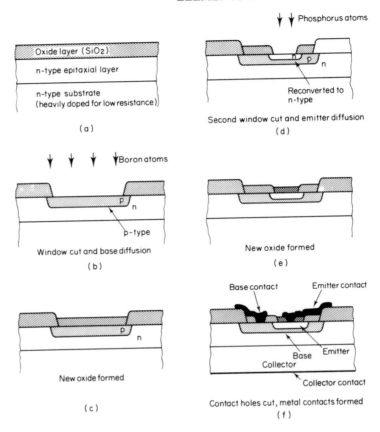

Figure 2.46. Some of the basic steps for the production of a monolithic, planar, diffused integrated circuit component (a npn bipolar transistor)

At this stage the devices or circuits on the wafer are tested, those which fail being marked with a dot; then the wafer is cut into individual units or *chips*. This is done by scribing the wafer with a diamond tool or partially cutting with a diamond saw and then breaking the wafer with a roller.

The final steps are to mount the chip on to a header, make electrical connections to the leads-out and encapsulate the chip into a package; a typical integrated circuit is shown in Figure 2.45. Electrical connections are made to the chip using thin (20–30 μm diameter) aluminium or gold wire; in some complex integrated circuits over 40 connections may be required, each having to be accurately positioned and correctly formed. Encapsulation is carried out to protect the chip from the effects of the atmosphere and dirt and is usually in the form of a plastic moulded package or hermetically sealed metal can. The latter, because of the high cost, tends to be used for high-reliability applications or devices with a high thermal dissipation.

The techniques described above can perhaps best be appreciated by considering the steps involved in producing an npn junction transistor (see Section 6.8.1). The steps are illustrated in Figure 2.46.

Initially an n-type epi-layer is grown on an n-type wafer and the surface of the epi-layer is then oxidized. A window is opened in the oxide layer and a p-type diffusion performed to convert some of the epi-layer from n-type to p-type, part of this region forms the base.

The surface in reoxidized and a new window opened. An n-type diffusion is now made to reconvert part of the p-region back to n-type to form the emitter region (Figure 2.46d). The surface is again reoxidized and new windows cut so as to expose the base and emitter regions so that metal can be evaporated on to the surface forming an electrical contact to these regions. The original epi-layer is the collector region and contact to it can be made via the substrate or through a window in the oxide as required.

Other devices which will be discussed in Chapter 6 have the following somewhat idealized cross-sectional structures (Figures 2.47 and 2.48).

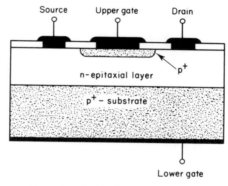

Figure 2.47. An n-channel junction field effect transistor (JFET) (the $+$ sign indicates very heavy doping)

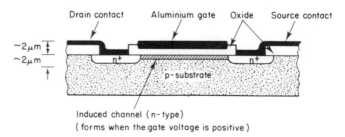

Figure 2.48. Induced n-channel, integrated circuit metal-oxide–silicon transistor (MOST)

The basic steps in the preparation of discrete devices and integrated circuits have been described above. There are, however, variations and additions to these basic steps that can be used to improve device operation for a particular purpose. For example, there are techniques to increase the power handling capacity, reduce contact resistance, increase the frequency of operation and isolate one device from another in integrated circuits.

2.12 SUMMARY

In this chapter the ordered atomic nature of crystalline solids has been discussed and techniques for producing large, high-quality crystals mentioned. The way in which semiconductor devices can be produced from such crystals as silicon and other semiconductors has been described, together with the actual structures of some representative devices. In the final chapter the operation of these and other devices will be described in terms of the models developed in the preceding chapters.

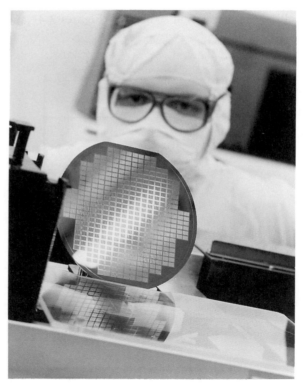

Figure 2.49. Silicon wafer inspection. Reproduced by permission of Fujitsu Microelectronics Limited

Figure 2.50. Natural crystals of lead sulfide, PbS, which has the NaCl crystal structure. (Photograph by B. Burleson)

PROBLEMS

2.1 What is the atomic radius in face centred and body centered cubic structures if the lattice constant is 1.0 nm?

2.2 Calculate the fraction of the volume of the unit cell actually occupied by atoms for the three cubic structures.

2.3 Copper has a density of 8.885×10^3 kg/m^3, a relative atomic mass of 63.57 and has a face centred cubic structure. Estimate:

    (a) How many moles of copper there are in 1 m^3.

    (b) How many atoms there are in 1 m^3.

    (c) The size of the unit cube.

    (d) The atomic radius of copper in the solid state.

    (e) The mass of the copper atom.

    (*Hint:* See the sample calculation on p. 87.)

2.4 Calculate the separations of the sets of planes which give rise to strong X-ray diffracted beams at angles θ of 3° and 9° in the first order, given that the X-ray wavelength is 0.1 nm.

2.5 At what angle will a diffracted beam emerge from the (110) planes of a face centred cubic crystal of unit cell length 0.5 nm? Assume diffraction occurs in the first order and that the X-ray wavelength is 0.4 nm.

2.6 Given that the separation S of a pair of lines on an X-ray powder photograph is 10 cm and that the camera radius is 24 cm calculate the separation of the planes responsible for the diffraction. Take the X-ray wavelength to be 0.16 nm.

2.7 An X-ray beam of wavelength 0.154 nm is incident on a set of planes of a silicon crystal. The first Bragg reflection is observed for an incidence angle of 34.5°. What is the plane separation and will there be any higher order reflections?

2.8 Draw the planes whose Miller indices are (421), (112), ($1\bar{1}0$) and ($\bar{1}21$) and state their intercepts on the x, y and z-axes.

2.9 The energy of interaction of two atoms separated by a distance r can be written as

$$E(r) = -\frac{\alpha}{r} + \frac{\beta}{r^8}$$

where α and β are constants.

(a) State which of the terms is 'attractive' and which is 'repulsive'.
(b) Show that for the particles to be in equilibrium $r = r_0 = (8\beta/\alpha)^{1/7}$.
(c) In stable equilibrium show that the energy of attraction is eight times that of the repulsion (in contrast to the forces of attraction and repulsion being equal).
(d) If the two atoms are pulled apart show that they will separate most easily when $r = (36\beta/\alpha)^{1/7}$.

2.10 Suppose that the interaction energy between two atoms is given by

$$E(r) = -\frac{A}{r^2} + \frac{B}{r^{10}}$$

Given that the atoms form a stable molecule with an internuclear distance of 0.3 nm and a dissociation energy of 4 eV calculate A and B.

Also calculate the force required to break the molecule and the critical distance between the nuclei for which this occurs.

2.11 Show that the Madelung constant for a one-dimensional array of ions of alternating sign with a distance a between the ions is $2\ln 2$.

2.12 How many processing steps would be required to fabricate a pn junction diode starting with an n^+ substrate?

3 Theories of Conduction and Magnetism

One of the reasons why the electrical properties of solids are so interesting can be seen immediately if the electrical resistivity of various materials is considered. Table 3.1 summarizes the results of a large number of measurements on a wide range of solids. The most striking feature is the enormous range of resistivity values which is covered, starting at $10^{-8}\,\Omega\,m$ and going up to $10^{20}\,\Omega\,m$; indeed, no other physical parameter covers such a wide range.

Table 3.1 Resistivities of some common materials at room temperature

Resistivity ($\Omega\,m$)	Material
10^{20}	Polystyrene
10^{18}	PTFE
10^{16}	Fused silica
10^{13}	Quartz
10^{12}	Glass
10^{11}	Diamond
10^{10}	Sodium chloride
10^{6}	Pure gallium arsenide
10^{4}	Ferrites
10^{3}	Pure silicon
10^{1}–10^{-6}	Doped silicon
10^{-1}	Pure germanium
10^{-2}	Selenium
10^{-7}	Lead, iron
10^{-8}	Copper, silver

Electrical resistivity therefore stands out as a most interesting parameter, and it is little wonder that so much time and effort has been spent in establishing and testing models of conduction in solids. When such models are set up, they must also account satisfactorily for the effects of other physical agencies on the electronic properties. They must explain why the resistance of a metal increases as the temperature rises, why the addition of small amounts of impurity to pure semiconductors improves their conductive properties and why the illumination of certain crystals with light of a suitable wavelength causes an increase in conductivity.

In this chapter, therefore, we shall examine some of the models which have been proposed and the extent to which they satisfy the conditions imposed by the experimental results.

3.1 CHARGE CARRIERS IN SOLIDS

The assumption that the electrical properties of solids are governed by the flow of electrons might at first glance appear to be perfectly reasonable. Before proceeding further, however, it would be instructive for us to consider a simple experiment that indicates the nature of charge carriers in a metal such as copper.

The nature of the charge carriers in metals was first investigated in 1917 by R. C. Tolman and T. D. Stewart, whose apparatus is shown schematically in Figure 3.1. Quite simply, a rotating coil of wire, connected in series with a ballistic galvanometer G, was suddenly stopped by means of the brakes. This caused a current to flow in the external circuit, and the direction of this current indicated that the carriers giving rise to it must have had a negative charge. Moreover, by measuring the coil rotational velocity, the deceleration time, the current flowing upon braking and the total length and resistance of the wire, the ratio of the charge to the mass of the carriers could be determined. Within the limits of experimental error, this value was found to agree with that obtained by Thomson for electrons. The explanation of the effect is simple. When the coil is rotating, both the metal ions and their loosely bound 'gas' of valence electrons (see Section 2.6.3) are moving at the same velocity. Upon braking, the motion of the atoms is arrested but the valence electrons continue to move, causing a current to flow in the external circuit and thereby giving a deflection on the galvanometer. Not surprisingly, this type of conduction is termed

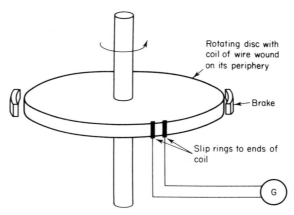

Rotating disc with coil of wire wound on its periphery

Brake

Slip rings to ends of coil

Figure 3.1. The Tolman–Stewart apparatus for measuring e/m for the current carriers in a metal

electronic conduction and it is found to occur in a large number of metals and semiconductors.

Another type of conduction process is found in ionic crystals that are heated to high temperatures. Experiments show that charge can be carried by positive or negative ions, which become more mobile as the temperature increases and thus more free to wander through the lattice when an electric field is applied across the crystal. Such a charge transfer process is termed *ionic conduction*. The mechanism whereby ions and indeed atoms move in a solid due to their thermal energy is called diffusion. There are various processes whereby this may happen; for example an atom may interchange places with another atom or it might occupy a vacant site (Section 2.3). This means that the motion of atoms or ions by diffusion is a step-like process, the elementary step length being a few lattice spacings. When all the elementary steps are added together, however, it is possible for an atom to cover quite large distances in the lattice.

Thus, even in the absence of an electric field, the atoms will always be in constant random diffusive motion. Other particles also move diffusively, for example electrons, and as will be seen in Chapter 6, electron diffusion is a crucial factor in the operation of junction devices. If there is a particle concentration gradient there will be a net flow of particles down the gradient, giving a diffusion current if the particles are charged (Section 5.5.1).

Two broad classes of conduction therefore exist—electronic and ionic. However, since it is metals and semiconductors that are of primary interest in this book, electronic conduction is our main concern.

3.2 CONDUCTION IN METALS—CLASSICAL APPROACH

The first attempt to explain electrical and thermal conduction in solids was provided by P. K. L. Drude, in 1900, who accepted the notion of a metal as a set of ions permeated by a 'sea' of electrons or an 'electron gas'. He suggested that the application of an electric field would accelerate the electrons in the field direction. When they encountered ions in the crystal lattice, they would be scattered, 'resistance' to the current flow would thus be created and the electrons would settle down at a constant *drift velocity* governed by the applied field strength.

The effect of the field on the motion of the electron gas is illustrated in Figure 3.2. Before the application of the field, the motion of the diffusing electrons is entirely random, so the average velocity in any specific direction is zero. When the field is applied, there is a general drift of the electrons towards the positive terminal. Although at any given instant any given electron might be travelling away from the positive terminal, the electron gas as a whole moves against the field. Of course, more electrons are supplied from the external source (for example a battery), so the process is a continuous one. Collisions still do

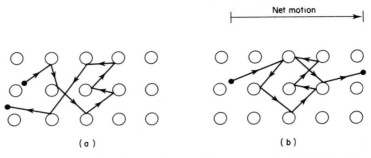

Figure 3.2. Electron motion: (a) random motion in zero electric field; (b) net drift in an applied electric field

occur, and the velocity between collisions — which can be compared to the thermal velocity of gas molecules — is obviously very much greater than the drift velocity of the gas as a whole. In copper, for example, the thermal velocity is some seven orders greater than the drift velocity (see Example 3.1).

Another simple mechanical analogy is the pin-table, where the pins represent the fixed ions and the metal balls the electrons. If a large number of metal balls is fired on to a flat table, they hit the pins, move off in random directions and give rise to no net motion in any given direction. However, if the table is tilted (equivalent to applying a field), there will be a drift of the balls as a whole down the table, although at any instant some might strike the pins in such a way as to move up the table.

Drude therefore used the kinetic theory of a neutral low-pressure gas to explain the behaviour of a dense electron gas, with very few modifications, together with Newton's laws. The assumptions of the theory are as follows:

(a) Between collisions the electrons move in straight lines and are not influenced by the heavy positive ions.
(b) The collisions are instantaneous events like rubber balls bouncing off a heavy fixed mass, which abruptly alter the velocity of the electrons (Figure 3.2b).
(c) The electrons lose all the extra energy gained from the applied field upon collision.
(d) The average distance an electron travels between collisions (the mean free path in kinetic theory) is of the order of the distance apart of the fixed ions (the lattice separation).

3.2.1   OHM'S LAW

The most firmly established experimental law relating to conduction in metals was Ohm's law, and Drude tested his theory by deriving this from first principles.

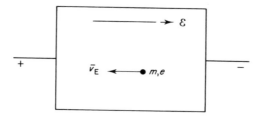

Figure 3.3. Moving charge in an electric field

Let $\mathcal{E}$ be the applied field, $\bar{v}_E$ the average velocity acquired by a given electron by acceleration against the field direction, $\bar{v}_T$ the average thermal velocity (equivalent to that for a classical gas) and e/m the ratio of the charge to the mass of the electron (Figure 3.3). When the field is applied, the force acting on an individual electron is $e\mathcal{E}$, so the acceleration a during the time between collisions is given by

$$a = \frac{e}{m}\mathcal{E}$$

If the average time of flight between collisions is denoted by τ, and according to assumption (c) the electrons start from rest after a collision, the average velocity in the field is

$$\bar{v}_E = \frac{e}{m}\mathcal{E}\tau$$

The drift velocity $\bar{v}_D$ is therefore made up of the average velocity due to the random thermal motion (which is zero in any specific direction) and the average velocity due to the field. The drift velocity $\bar{v}_D$ therefore equals the average field velocity $\bar{v}_E$, so

$$\bar{v}_D = \frac{e}{m}\mathcal{E}\tau \tag{3.1}$$

If there is a total of n electrons per unit volume taking part in the conduction process, the total current density, that is the current flowing over unit area, will be the total charge density times the drift velocity. Thus

$$J = ne\bar{v}_D \tag{3.2}$$

Substituting the value of $\bar{v}_D$ from equation (3.1) into equation (3.2) gives

$$J = \frac{ne^2\tau}{m}\mathcal{E} \tag{3.3}$$

Although perhaps not immediately obvious, this is recognizable as Ohm's law, because if we write the conductivity as

$$\sigma = \frac{ne^2\tau}{m} \tag{3.4}$$

and recall that the resistivity ρ is given by

$$\rho = \frac{1}{\sigma} = \frac{m}{ne^2\tau} \tag{3.5}$$

and that for a wire of cross-sectional area A, length l with resistance R, the resistivity is given by

$$\rho = \frac{RA}{l}$$

equation (3.3) can be rewritten as the familiar $V = RI$, since

$$\varepsilon = \frac{V}{l} \qquad \text{and} \qquad J = \frac{I}{A}$$

Moreover, we can also write equation (3.4) as

$$\sigma = ne\mu \tag{3.6}$$

where

$$\mu = \frac{e\tau}{m}$$

This parameter, called the mobility, is defined as the average drift velocity per unit electric field, that is

$$\mu = \frac{\bar{v}_D}{\varepsilon}$$

As already mentioned, the thermal velocity is much larger than the drift velocity. An estimate of $\bar{v}_T$ can be made easily from kinetic theory which states that

$$\tfrac{1}{2}m\bar{v}_T^2 = \tfrac{3}{2}kT$$

where k is Boltzmann's constant.

Example 3.1

Compare the value of the average thermal velocity with the drift velocity of an electron gas at 300 K subjected to an electric field of 10 V m^{-1}. The intercollision time is 10^{-14} s.

Thermal velocity: $\frac{1}{2}m\bar{v}_T^2 = \frac{3}{2}kT$

$$\bar{v}_T = \sqrt{3kT/m}$$

$$\bar{v}_T = \sqrt{\frac{3 \times (1.38 \times 10^{-23})\,(300)}{9.1 \times 10^{-31}}}$$

$$\bar{v}_T = 1.2 \times 10^5 \text{ m s}^{-1}$$

Drift Velocity:

$$\bar{v}_D = \frac{e}{m}\,\mathcal{E}\tau$$

$$\bar{v}_D = \frac{(1.6 \times 10^{-19})\,(10)\,(10^{-14})}{9.1 \times 10^{-31}}$$

$$\bar{v}_D = 17.6 \times 10^{-3} \text{ m s}^{-1}$$

All the parameters in equation (3.4) can be determined independently and substituting for n, e, m and τ gives values of σ that agree very well with those obtained by direct measurement.

3.2.2 WIEDEMANN–FRANZ LAW

Implicit in the Drude theory is the assumption that electrons are not only the agencies of electrical conduction but also are responsible for the transport of thermal energy in a solid. An empirical relationship, called the Wiedemann–Franz law, had been established in 1853, which relates the thermal and electrical conductivities by the expression

$$\frac{K}{\sigma} = C_{WF}\,T$$

where K is the thermal conductivity, T the absolute temperature and C_{WF} has the nature of a universal constant which has been determined accurately for metals and found to have an average value of $2.31 \times 10^{-8} \text{ W}\,\Omega\,\text{K}^{-2}$.

Drude attempted to derive a theoretical expression for C_{WF} by applying the kinetic theory of gases to the random thermal motion of the electrons in a metal as follows. Consider a region filled with an electron gas, with n electrons per unit volume. Let the average thermal velocity be $\bar{v}_T$, the mean free path be λ, and let a thermal gradient exist along the x direction as illustrated in Figure 3.4. Suppose that the average energy of a given electron crossing a given plane is E, so the energy at planes a distance λ away will differ from E by an amount $\lambda\,dE/dx$.

From kinetic theory, the number of particles flowing in a given direction is $\frac{1}{6}n\bar{v}_T$, so the net energy flow across the central plane per unit time will be

$$\frac{1}{6}n\bar{v}_T \times 2 \frac{dE}{dx} = \frac{n\bar{v}_T}{3}\frac{dE}{dx}$$

From the normal theory of heat flow, this can be equated to the product of the thermal conductivity K and the temperature gradient. Thus,

$$\frac{n\bar{v}_T}{3}\frac{dE}{dx} = K\frac{dT}{dx}$$

Now

$$n\frac{dE}{dx} = n\frac{dE}{dT}\frac{dT}{dx} = C_v\frac{dT}{dx}$$

Combining the previous two equations,

$$K = \tfrac{1}{3}C_v\,\bar{v}_T\lambda \tag{3.7}$$

where C_v is the specific heat capacity at constant volume. The mean free path λ can be written in terms of the average time between collisions and the average thermal velocity; that is $\lambda = \tau\bar{v}_T$. From equations (3.4) and (3.7) therefore,

$$\frac{K}{\sigma} = \frac{1}{3}\frac{C_v\bar{v}_T\lambda m}{ne^2\tau} = \frac{1}{3}\frac{C_v m\bar{v}_T^2}{ne^2}$$

However, from kinetic theory, $C_v = \tfrac{3}{2}nk$ and $m\bar{v}_T^2$ is twice the thermal kinetic energy (or $2 \times \tfrac{3}{2}kT$), where k is Boltzmann's constant. Therefore,

$$\frac{K}{\sigma} = \frac{3}{2}\left(\frac{k}{e}\right)^2 T$$

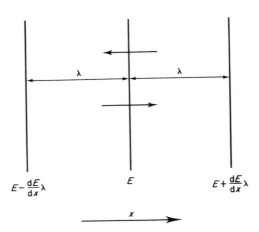

Figure 3.4. Energy of electrons in a region filled with an electron gas

Table 3.2 Electrical and thermal conductivities measured at 293 K

Metal	$\sigma \times 10^7$ $(\Omega^{-1} m^{-1})$	K $(W\,m^{-1}\,K^{-1})$	$C_{WF} \times 10^{-8}$ $(W\,\Omega\,K^{-2})$
Silver	6.15	423	2.45
Copper	5.82	387	2.37
Aluminium	3.55	210	2.02
Sodium	2.10	135	2.18
Cadmium	1.30	102	2.64
Iron	1.00	67	2.31
Lead	0.45	34	2.56

Substituting values of k and e, the theoretical Wiedemann–Franz constant is found to be $1.22 \times 10^{-8}\,W\,\Omega\,K^{-2}$, which although approximately half the average experimental value is nevertheless of the same order and quite acceptable considering the simplicity of the model and the nature of the assumptions made. Indeed, it may be seen from the values of electrical and thermal conductivity given in Table 3.2 that good electrical conductors are also good thermal conductors, which clearly supports Drude's theory.

The successes of the theory were therefore considerable and it may be considered as the 'Bohr' equivalent for solids. Although the Drude theory provides a conceptually easy model there are a number of instances where it proves inadequate and we shall now review some of these.

3.3 BREAKDOWN OF THE CLASSICAL THEORY OF CONDUCTION

3.3.1 MEAN FREE PATHS

One of Drude's fundamental assumptions was that the mean free path should be of the same order as the lattice spacing. Clearly, any true test of the theory must involve a determination of λ. It is possible to calculate the mean free path by determining τ from equation (3.3), and then use the expression

$$\lambda = \tau \bar{v}_T$$

In a metal such as copper, according to the Drude theory, each atom should contribute one electron to the conduction process. The concentration n of conduction electrons is then the number of copper atoms per unit volume, which is Avogadro's constant divided by the molar volume. The molar volume is the molecular weight divided by the density of copper. The number of conduction electrons is then

$$n = \frac{6.02 \times 10^{23} \times 10^3 \times 8.93 \times 10^3}{63.54} = 8.5 \times 10^{28}\,m^{-3}$$

Table 3.3 Mean free paths of various metals at 300 K

Metal	λ (nm)
Lithium	11
Sodium	35
Potassium	37
Copper	53
Silver	57

Substituting this value and the other constants into equation (3.3) gives $\tau = 2.4 \times 10^{-14}$ s; then taking $\bar{v}_T = 1.2 \times 10^5$ m s^{-1} gives $\lambda = 2.9$ nm for copper. Table 3.3 shows some observed mean free paths for various metals at 300 K, from which we see that these are very much larger than those supposed by Drude.

3.3.2 SPECIFIC HEATS

Another serious objection to the theory is its failure to account for the observed specific heat capacity of metals, which classically should equal $\frac{9}{2}nk$ per unit volume ($\frac{3}{2}nk$ for each of the lattice kinetic and potential energies plus $\frac{3}{2}nk$ for the electrons, assuming one valence electron per atom). However, the measured specific heat capacities at room temperature are approximately $3nk$ per unit volume, indicating that very few of the available conduction electrons make any contribution to the specific heat capacity (see example 5.1).

This discrepancy can only be accounted for by supposing that the value of n used by Drude is incorrect, implying that only a certain small fraction of the electrons in the electron gas should be considered. The theory is also unable to account for resistivity variations due to changes in temperature and pressure and for the effect of impurities on the conductivity.

3.3.3 THE HALL EFFECT

In 1879 G. Hall performed an experiment, the subsequent results of which seem to contradict completely the classical picture of conduction. If a sample of conducting material is placed in a uniform magnetic field and a current is passed along the length of the specimen, as shown in Figure 3.5, it is found that a voltage develops at right angles to both the direction of the current flow and that of the magnetic field. This voltage is known as the Hall voltage, and its value is found to depend on the magnetic field strength and on the current passed. We can understand why the Hall voltage arises quite simply by considering that the current represents a beam of negatively charged electrons moving at right angles to the magnetic field. The beam is therefore deflected, so one side of the specimen acquires a negative charge thereby establishing a potential difference across the sample. The mathematics of the Hall effect can be based on the simple dynamics of charges moving in electromagnetic fields.

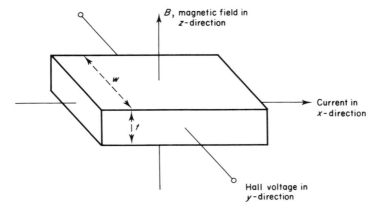

Figure 3.5. The Hall effect

Consider a slab of material as shown in Figure 3.5 subjected to a magnetic field perpendicular to the direction of electron current flow. Owing to the magnetic field B, a force F_B will be experienced by the electron beam such that

$$F_B = Bev_x$$

where v_x is the velocity of the electrons in the x direction. The electron beam is therefore deflected so that, when equilibrium is established and no further current flows in the y direction, a field $\mathcal{E}_H$ acts in the y direction. An additional force will therefore act on each electron given by

$$F_E = e\mathcal{E}_H$$

In equilibrium a steady state has been set up and the electrical force F_E and the magnetic force F_B must be equal and opposite; that is,

$$e\mathcal{E}_H = Bev_x$$

The Hall voltage V_H will then be

$$V_H = \mathcal{E}_H w = Bv_x w \tag{3.8}$$

where w is the width of the sample. The current density J_x is given by the equation

$$J_x = nev_x = \frac{I_x}{wt} \tag{3.9}$$

where t is the sample thickness and I_x is the current in the x direction. Combining equations (3.8) and (3.9) gives

$$\frac{V_H t}{BI_x} = \frac{1}{ne} = R_H \tag{3.10}$$

where R_H is defined as the Hall coefficient. Moreover, since $\sigma = ne\mu$, there follows the simple relationship:

$$\mu = R_H \sigma \qquad (3.11)$$

Therefore, in a given experiment, if the Hall constant R_H and the conductivity σ are measured, then the mobility μ can be found from equation (3.11) as well as the carrier concentration n.

The results of some Hall effect measurements at room temperature on various conductors are shown in Table 3.4. There is clearly a wide range of Hall coefficient magnitudes and hence of mobilities. However, the most significant feature is that Hall constants of both signs occur and, bearing in mind equation (3.10), this can only imply that conduction by electrons will give a negative Hall coefficient and conversely a positive coefficient implies that the carriers are positive. This seems to contradict the fundamental finding of Tolman's experiment that the charge transfer mechanism in solids involves the motion of electrons.

Example 3.2

A semiconducting sample is 5 mm long, 4 mm wide and 2 mm thick, and when a 2 V supply is connected a current of 40 mA flows along the specimen. As a result a Hall voltage of 15 mV develops across the width of the specimen, when a magnetic field of 0.1 T is applied. Determine the carrier density and mobility.

$$\mathcal{E}_H = V_H / w = (15 \times 10^{-3})/(4 \times 10^{-3})$$
$$= 3.75 \text{ V m}^{-1}$$
$$J_x = I_x / wt = (40 \times 10^{-3})/(4 \times 2 \times 10^{-6})$$
$$= 5 \times 10^3 \text{ A m}^{-2}$$
$$R_H = V_H t / BI_x = \mathcal{E}_H / J_x B$$
$$= (3.75)/(5 \times 10^3) \times (0.1)$$
$$= 7.5 \times 10^{-3}$$
$$n = 1/R_H e = 1/(7.5 \times 10^{-3})(1.6 \times 10^{-19})$$
$$= 8.3 \times 10^{20} \text{ m}^{-3}$$

Now $\rho = VA/Il = 2 \times (2 \times 4 \times 10^{-6})/(40 \times 10^{-3})(5 \times 10^{-3})$
$$= 0.08 \ \Omega^{-1} \text{m}^{-1}$$

and $\qquad \mu = R_H \sigma = R_H / \rho = 7.5 \times 10^{-3}/0.08$
$$= 0.938 \text{ m}^2 \text{V}^{-1} \text{s}^{-1}$$

Before leaving the Hall effect, it should be pointed out that the direct proportionality between the Hall voltage and the applied magnetic field for a given current is used to advantage in Hall probe magnetometers, which provide a direct and accurate way of measuring magnetic fields.

Table 3.4 Hall coefficients and mobilities for some metals at 300 K

Metal	$R_H \times 10^{-10}$ (m^3 C^{-1})	μ (m^2 V^{-1} s^{-1})
Silver	-0.84	0.0056
Copper	-0.55	0.0032
Gold	-0.72	0.0030
Sodium	-2.50	0.0053
Zinc	$+0.30$	0.0060
Cadmium	$+0.60$	0.0080
Magnesium	-0.92	0.0021

3.3.4 ADDITIVE NATURE OF RESISTIVITY

As shown by the Drude theory, the main source of the resistivity of a metal is the scattering of the electrons by the ions. The higher the temperature, the more vigorous the vibration of the ions, the greater the scattering (the *phonon* scattering, described in Chapter 1 and in Section 2.8.4) and hence the greater the resistivity.

However, a contribution to the scattering is also provided by impurities, such as 'foreign' atoms in either interstitial or substitutional positions or by mechanical deformations which can produce a range of crystal lattice defects (see Section 2.8). A long-established experimental observation was that the contributions to the total resistivity from these various sources were additive. This rule, known as *Mathiessen's law*, can be written as

$$\rho_{Total} = \rho_{Lattice} + \rho_{Impurities}$$

Noting that $\rho = 1/\sigma$, and using equation (3.5), the total resistivity can be written as

$$\rho_{Total} = \frac{m}{ne^2} \left(\frac{1}{\tau_L} + \frac{1}{\tau_I} \right)$$

or

$$\rho_{Total} = \frac{m\bar{v}_T}{ne^2} \left(\frac{1}{\lambda_L} + \frac{1}{\lambda_I} \right) \tag{3.12}$$

where the symbols are as defined in Section 3.2.1.

The parameters characterizing the processes are the lattice scattering mean free path λ_L and the impurity scattering mean free path λ_I, of which only the former is temperature dependent. The impurity scattering is therefore only of importance at low temperatures, when the lattice scattering has virtually reduced to zero (see also p. 157). A knowledge of the effects on resistivity of the addition of small amounts of impurities is of particular importance in alloys. For example, the addition of about 3% nickel to otherwise pure copper increases its resistivity by a factor 4 at room temperature.

Example 3.3

A pure copper specimen has a resistivity of $1.72 \times 10^{-8}\,\Omega$ m and a mean free path of 40 nm at 300 K. Estimate the resistivity of copper containing 1% Ni if the mean free path for collisions with the impurity atoms is 60 nm. The average thermal velocity at 300 K is 1.2×10^5 m s^{-1}.

Solution. $\tau = \lambda / \bar{v}_T$

$$\tau_L = \frac{40 \times 10^{-9}}{1.2 \times 10^5} = 3.3 \times 10^{-13}\,\text{s}$$

$$\tau_I = \frac{60 \times 10^{-9}}{1.2 \times 10^5} = 5 \times 10^{-13}\,\text{s}$$

$$\sigma_L = \frac{1}{\rho_L} = \frac{ne^2}{m}\tau_L$$

$$n = \frac{m}{\rho_L e^2 \tau_L} = \frac{9 \times 10^{-31}}{(1.72 \times 10^{-8})(1.6 \times 10^{-19})^2(3.3 \times 10^{-13})}$$

$$= 6.3 \times 10^{27}\,\text{m}^{-3}$$

Now $\rho_{\text{Total}} = \rho_L + \rho_I$

$$\rho_I = \frac{m}{ne^2 \tau_I} = \frac{9 \times 10^{-31}}{(6.3 \times 10^{27})(1.6 \times 10^{-19})^2(5 \times 10^{-13})}$$

$$\rho_I = 1.12 \times 10^{-8}\,\Omega\,\text{m} \quad (1\%\ \text{Ni})$$

$$\rho_{\text{Total}} = 1.72 \times 10^{-8} + 1.12 \times 10^{-8}$$

$$= 2.84 \times 10^{-8}\,\Omega\,\text{m}$$

3.4   THE MAGNETIC PROPERTIES OF SOLIDS

The previous section presented a model of the electrical and thermal conduction in metals such as copper, silver and gold in terms of the behaviour of electrons. However, there are other materials, which although not exhibiting particularly good conductive properties are nevertheless interesting from another point of view, namely they are magnetic. It is again models based on the behaviour of electrons which are used to explain such behaviour.

Magnetism is perhaps that aspect of solid state physics which has been familiar to man for the longest period of time. The ability of the lodestone (magnetite) to attract iron objects was known 3000 years ago to the ancient Greeks, and long before any conclusive theories of magnetism had been drawn up a wealth of experimental data on the phenomenon was available. For example, it was known that iron could be magnetized by stroking it with another magnet, by hammering it whilst lying in the magnetic meridian or simply by holding it somewhere in the vicinity of a strong magnet. In fact, almost everything known about the magnetic properties of materials has been derived from experimental discoveries and from a few inspired guesses. This is undoubtedly due to the fact that a quantitative description of magnetism requires the application of quantum mechanics and electromagnetic theory to many atom systems — which invariably proves too complex to allow reasoning to be made from the general to the particular.

This chapter, therefore, can only 'scratch at the surface' of the subject and present a simplified description of the atomic processes giving rise to the observed phenomena.

The three basic facts concerning magnetic properties that require explanation are:

(a) Some materials are magnetic, even in the absence of any applied magnetic field, and become more magnetic when a very weak magnetic field is applied to them.
(b) Some materials lose their initially strong magnetism when heated above a certain critical temperature and become comparatively weakly magnetized.
(c) In some materials an induced magnetism in a direction opposite to that of an externally applied field can be detected experimentally.

3.4.1 ATOMIC THEORY OF MAGNETISM

The first attempt to explain magnetism on an atomic scale was made by J. A. Ewing, who supposed that the molecules making up the material themselves behaved as tiny bar magnets. He assumed that, in an unmagnetized specimen, these molecular magnets were so arranged in small groups that no resultant magnetic field arose (Figure 3.6a). In a magnetized specimen, all the molecules would have their magnetic moments pointing in the same direction, giving rise to a magnetic field (Figure 3.6b). Ewing's theory was fairly close to the truth, but his assumption that the molecules were the elementary magnets had to be discarded. It is now known that the fundamental magnetic units are the *extranuclear electrons*.

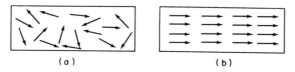

(a) (b)

Figure 3.6. Ewing's theory of the arrangement of molecules: (a) in an unmagnetized specimen; (b) in a magnetized specimen

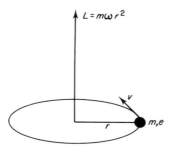

Figure 3.7. Angular momentum of an orbiting electron

Consider an electron of mass m, describing an orbit of radius r with velocity v and angular velocity ω, as shown in Figure 3.7. From the simple ideas of current electricity, a current I can be regarded as a charge dq moving at a velocity $v = ds/dt$. Since $I = dq/dt$,

$$I\,ds = \frac{dq}{dt}\,ds = dq\,\frac{ds}{dt}$$

Putting $dq = e$ (the electronic charge) gives

$$I\,ds = ev$$

where v is the linear velocity. However, for a circular orbit,

$$ds = 2\pi r$$

so

$$I \times 2\pi r = ev$$

and

$$v = r\omega$$

and thus

$$I = \frac{e\omega}{2\pi}$$

From electromagnetic theory, the magnetic moment of a closed circuit of current I with cross-sectional area A is IA.* The magnitude of the magnetic moment m produced by the circular motion of the electron is therefore given by

$$\mu_e = \tfrac{1}{2}e\omega r^2 \tag{3.13}$$

---

*Some authors define the magnetic moment by $\mu_0 IA$ (Kennelly notation), whereas the notation here is that of Sommerfield, which is now recommended.

The angular momentum of the circular motion, which may be represented as a vector normal to the plane of the orbit, is given by

$$L = m\omega r^2$$

where L is the magnitude of vector $\mathbf{L}$. Equation (3.13) may be rewritten as

$$\mu_e = \frac{e}{2m}\,\mathbf{L} \tag{3.14}$$

where μ_e is also a vector antiparallel to $\mathbf{L}$.

So far, in this working, purely classical reasoning has been used, similar to that employed for the Bohr theory (see Section 1.8). However, quantum mechanics must be applied to this problem, the principal result being that the electron may only move in orbits where its angular momenta are integral multiples of $h/2\pi$, that is

$$L = l\,\frac{h}{2\pi}$$

Combining this with equation (3.14)

$$\mu_e = -e\,\frac{h}{4\pi m}$$

This equation indicates that the magnetic moment resulting from the orbital motion can change its value only in units of

$$\mu_B = \frac{he}{4\pi m}$$

This is the smallest theoretically possible magnetic moment and is called the *Bohr magneton*. Substituting the values of the known constants gives

$$\mu_B = 9.27 \times 10^{-24}\ \text{A m}^2$$

In the quantum theory of the atom, the effect of this orbital magnetic moment is the introduction of the third quantum number m_l, which relates to the number of ways an electron having a given n and l value can orientate itself in an external magnetic field. Quantum theory shows that the angular momentum vector, and hence the magnetic moment, can only assume certain prescribed directions in space.

The electron moving round the nucleus is equivalent to a small circulating electric current, which, by the normal laws of electrodynamics, produces a magnetic field. As can be seen in Figure 3.8, the field is similar to that produced by a bar magnet, whose strength can be expressed in terms of its magnetic moment.

If an external magnetic field is applied to an atom, the direction of the angular momentum vector, and hence the magnetic moment, precesses about the field

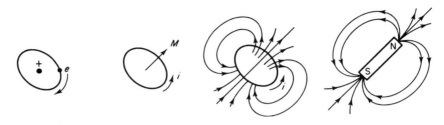

Figure 3.8. Representation of an electron in its orbit by a bar magnet

direction. The situation is then very similar to that of a gyroscope or top which precesses about the direction of the earth's gravitational field, as illustrated in Figure 3.9.

The major difference between the classical and the quantum mechanical situations is that, whereas the axis of rotation of the top may set at any angle θ to the field depending on its various dynamic parameters, a 'quantum mechanical top' will take on only those values that make the magnetic quantum number $m_l = 0, \pm 1, \pm 2, \ldots, \pm l$. For example, if $l = 2$, the electron's magnetic moment can take up the special orientations relative to the external field shown in Figure 3.10. It can now be seen that the magnetic quantum number is the projected component of the angular momentum L on the field direction.

What now is the significance of the fourth quantum number—the spin quantum number? It is supposed that, in addition to its orbital motion, the electron also spins about its own axis and thereby has its own angular momentum vector s. According to whether the electron is spinning clockwise or anticlockwise, the angular momentum vector points either upwards or downwards. As with the orbital motion, a magnetic moment is assumed to be associated with this spinning motion. If an external magnetic field is applied to a spinning electron, only two acceptable orientations are found to be possible,

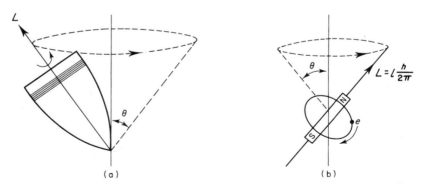

Figure 3.9. Precession: (a) of a classical top; (b) of an electron orbit

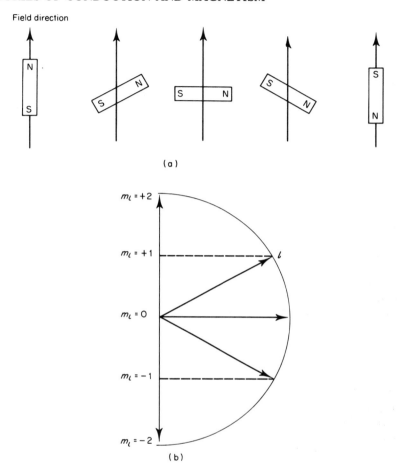

Figure 3.10. Possible orientations of the electron orbit (bar magnet) to the direction of the applied magnetic field

oppositely directed along the axis of rotation of the electron about itself. The projections along the field direction in this case are $+\frac{1}{2}$ and $-\frac{1}{2}$, again in units of $h/2\pi$. These then are the spin quantum numbers m_s referred to in Chapter 1 and illustrated in Figure 3.11.

It should be emphasized that the concept of spin is purely quantum mechanical and that the model introduced here is simply an attempt to present a more visual idea of what is taking place within an atom. In other words, the notion of a spinning electron is a classical description of a quantum mechanical process

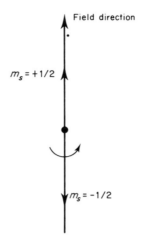

Figure 3.11. Spin quantization

and should not be taken too literally. The relationship between the magnetic moment and the angular momentum due to spin is given by

$$\mu_e = \frac{e}{m}s$$

Comparing this with equation (3.14) shows that the proportionality factors for orbital and spin motion differ by a factor of two.

So far, only a single electron orbit has been considered. For more complex atoms, in which large numbers of electrons are present, the total effect of all the electron orbital motions and all the electron spins must be taken into account in order to predict the overall magnetic moment associated with a given atom. Methods are available for summing the orbital and spin angular momenta, and hence deriving the total magnetic moment. Certain experimental results show that the orbital motion usually makes little or no contribution to the total angular momentum in crystalline solids. In the following discussions, therefore, only the magnetic moment due to spin need be considered.

Whether or not a given atom has a permanent magnetic moment depends on how the electrons are arranged around the nucleus. In Section 1.8 we saw how the Periodic Table can be built up using the Pauli principle together with the basic postulate of physics that states of lower energy are preferred to higher energy states. Moreover, according to the so-called Hund rule, states with $m_s = +\frac{1}{2}$ should be filled before those with $m_s = -\frac{1}{2}$. For descriptive purposes, the two spin states are distinguished by the terms 'spin-up' and 'spin-down' and are represented diagrammatically as small arrows pointing upwards (↑) or downwards (↓). If two spins point in opposite directions, they are said to be *antiparallel* (↑↓); if they point in the same direction they are said to be *parallel* (↑↑). If these ideas are applied to certain elements in the periodic system, it can at once be appreciated why magnetic properties are associated with them. A representative set of elements is shown in Table 3.5.

Let us consider an atom of oxygen. The first available electron is placed in the 1s spin-up state, the second in the 1s spin-down state. This process is repeated through the 2s subshell until the 2p subshell is reached. Three of the remaining four available electrons are placed in the 2p spin-up states and the last one in the first available 2p spin-down state. Since six electrons are required to complete a p subshell, two states with parallel spins remain empty.

The oppositely directed spins in the full s subshells exactly counteract one another, giving rise to no resultant spin momentum and hence no magnetic moment. On the other hand, the two upward pointing uncompensated spins in the 2p subshell add to give a resultant magnetic moment. This spin pairing, or compensation, is the key to why certain atoms have magnetic moments; completely filled shells do not contribute to the magnetic moment, the contribution coming only from incomplete shells. Thus while oxygen will have a net magnetic moment, a neon atom, with all its shells complete, should show no magnetic effects. The very strong magnetic effects associated with iron also result from unfilled shells, but other factors, discussed in Section 3.5.3, are also important.

At first, it might appear strange that the 4s levels in the iron atom should be filled before all the 3d levels, but the experimental evidence indicates that certain of the latter are energetically higher than either of the 4s levels. Thus for iron, which has six electrons in the 3d subshell, we would expect a total of four Bohr magnetons for the magnetic moment (five with 'up' spin and one with 'down' spin). Measurements show, however, that the effective magnetic moment of iron is 3.2, which is not too surprising if we recall that in the solid the atomic levels are broadened into bands, as discussed in Chapter 4, thereby invalidating the simple atomic picture. The non-integral number of Bohr magnetons has been explained on the basis of an overlapping between the broad 4s band and the narrow 3d band. Nevertheless, a permanent magnetic moment is associated with iron.

The Periodic Table thus indicates why certain materials should show magnetic effects, but does not account for their detailed behaviour or answer the questions posed in the introduction to this section. A complete explanation requires knowledge of the arrangement and distribution of the elementary electron 'bar magnets' within the solid.

3.5 CLASSIFICATION OF TYPES OF MAGNETISM

The magnetic moment per unit volume of a magnetic substance is called the intensity of magnetization and is denoted by a vector **M**. The unit of **M** is the

Table 3.5. The spin states of three representative elements

Element	K shell	L shell		M shell			N shell	Expected magnetic moment (Bohr magnetons)
	1s subshell	2s subshell	2p subshell	3s subshell	3p subshell	3d subshell	4s subshell	
O	↑↓	↑↓	↑↑ ↓					2
Ne	↑↓	↑↓	↑↑↑ ↓↓↓					0
Fe	↑↓	↑↓	↑↑↑ ↓↓↓	↑↓	↑↑↑ ↓↓↓	↑↑↑↑↑ ↓	↑↓	4

ampere per metre. The magnetic induction or magnetic flux density **B** is related to **M** by the expression

$$\mathbf{B} = \mu_0(\mathbf{M} + \mathbf{H}) \tag{3.15}$$

where **H** is the magnetic field and the relationship between intensity of magnetization and magnetic field can be expressed by

$$\mathbf{M} = \chi_m \mathbf{H} \tag{3.16}$$

where χ_m is the magnetic susceptibility. The relationship between magnetic field and magnetic flux density is

$$\mathbf{B} = \mu \mathbf{H} \tag{3.17}$$

where μ is the permeability. However, the relative permeability μ_r, which is given by

$$\mu_r = \frac{\mu}{\mu_0}$$

is normally used, and hence according to equations (3.15) to (3.17)

$$\mu_r = 1 + \chi_m \tag{3.18}$$

Magnetic materials can be classified according to their χ_m value and the way in which this varies with magnetic field intensity and temperature. The categories into which magnetic substances can be divided are diamagnetism, paramagnetism, ferromagnetism, antiferromagnetism and ferrimagnetism, and we shall now consider these briefly in turn.

3.5.1 DIAMAGNETISM

Diamagnetism is an exceedingly weak magnetism exhibited by all materials whereby a magnetization opposite to the field direction is induced. The relative susceptibility is negative, temperature independent and of the order of 10^{-5}.

To explain diamagnetism, consider the orbital motion of the electrons about the nucleus, which, as has been shown, can be approximated to a current-carrying coil. If this is placed in a magnetic field, it will turn and try to set at right angles to the latter, but in so doing a current will be induced in the coil. By Lenz's law, this induced current will oppose the original current and therefore give rise to a magnetic moment oppositely directed to the applied field. If the resistance of the coil is assumed to be zero — which is a good approximation to an orbiting electron — the induced current will persist as long as the applied field is maintained.

It can be shown that the diamagnetic susceptibility per unit volume is given by the expression

$$\chi_m^{d(v)} = -\left(\frac{\mu_0 e^2}{6m}\right)\{NZ \langle r \rangle^2\}$$

where N is the number of atoms in unit volume of the solid, Z is the atomic number, and $\langle r \rangle^2$ is the mean square distance of the electrons from the nucleus. For normal solids, $N \approx 5 \times 10^{28} \, \text{m}^{-3}$ and $\langle r \rangle^2 \approx 10^{-20} \, \text{m}^2$, so that, substituting in the above expression, the diamagnetic susceptibility reduces to

$$\chi_m^{d(v)} \approx - Z \times 10^{-7}$$

The temperature invariance of $\chi_m^{d(v)}$ can be readily appreciated if we recall that diamagnetism is caused predominantly by electrons moving in orbits lying deep within the atom, which are unaffected by thermal agitation. While diamagnetism occurs for all elements it is usually too weak to be detected, and for elements with incomplete shells and associated permanent magnetic moments, the magnetic effects are usually so intense that the diamagnetic effect can be neglected.

3.5.2   PARAMAGNETISM

Atoms with unpaired spins are said to be paramagnetic, but normally members of the transition group (including iron, cobalt and nickel) are treated separately under the heading of ferromagnetism (Section 3.5.3). Paramagnetic materials include manganese, platinum, tungsten and some members of the rare earth group (for example dysprosium and erbium), of which some are also ferromagnetic (for example, gadolinium). In addition, ions formed by removing or adding electrons to basic atoms (thereby creating unpaired spins) are also paramagnetic.

The main characteristic of a paramagnetic material is that a magnetic moment may be induced in the specimen under the action of an externally applied field; the elementary magnets align themselves in the same direction as the field. This is explained by the fact that paramagnetism is associated with the outermost electrons, which are less tightly bound to the atom than the inner electrons causing diamagnetism. The permanent magnetic moments of the individual atoms are therefore able to change their direction freely and, in the absence of an applied field, are randomly directed owing to thermal agitation. When the field is applied, the magnetic moments tend to become parallel to the field direction and overcome to some extent the disrupting influence of temperature. As soon as the field is removed, they are restored again to random positions.

It has been found experimentally that the paramagnetic susceptibility χ_m, which lies in the range 10^{-3}–10^{-5}, is inversely proportional to the absolute temperature, in accordance with the so-called Curie law ($\chi_m^p = C_c/T$), where C_c is the *Curie constant* ($C_c = N\mu_B^2/3k$).

3.5.3   FERROMAGNETISM

When we talk about magnets, we are usually referring to a ferromagnet such as a piece of iron that has the ability to pick up other pieces of iron. Such a material is said to be a permanent magnet, since it exhibits magnetic properties even in the absence of an applied field. It is found experimentally, however, that the magnetization is increased when an external magnetic field is applied to the specimen. Moreover, the relationship between magnetization **M** and applied field **H** is non-linear and **M** rises to a certain saturation value. If the applied field is then decreased gradually, the magnetization does not return to the same value for the same field, so a hysteresis effect occurs, as illustrated in Figure 3.12. The strong magnetism is also found to disappear at some critical value of temperature; the specimen becomes demagnetized, behaves like a paramagnet and only shows strong magnetic effects again when an external field is applied.

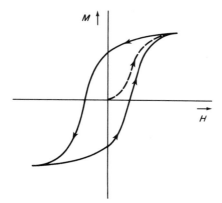

Figure 3.12. Hysteresis loop for a ferromagnetic material

These then are the experimental facts that any explanation of ferromagnetism must take into account. The modern theory of ferromagnetism, due to W. Heisenberg and P. Weiss, is an extension of that put forward by Ewing. The main suppositions are:

(a) Ferromagnetic materials are made up of large numbers of *domains*, which can be smaller than the grains or crystals of the metal itself. Each domain contains about 10^9–10^{15} atoms.
(b) Within a given domain, all the spins are aligned in the same direction, so the resulting magnetization of a given domain is the maximum possible value for a given material and temperature.

If a slice were taken through a ferromagnetic specimen, the domain arrangement could be as shown schematically in Figure 3.13. Not all domains have the same size, and very often the actual magnetization of the specimen as a whole is governed by that of the largest domain. The boundaries between the domains

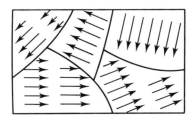

Figure 3.13. Ferromagnetic domains

represent regions where the direction of the spins of the constituent electrons change. This latter point will be discussed in more detail below.

Perhaps the most puzzling feature of this theory is that all the spins are aligned within a given domain. Weiss proposed that this results from a permanent intramolecular field which extends throughout the volume of a given domain. The theory of this field can be derived using quantum mechanics. The electrons whose spins are responsible for the magnetic moments of the individual atoms are in the M shell, which as we see from Table 3.5 is incomplete for iron. The interatomic spacings, however, are such that the orbits of these electrons can interpenetrate one another, and in this way an interaction takes place that gives rise to the intramolecular field. The interaction is conveniently described in terms of an exchange integral which is denoted by the symbol $\mathscr{J}$, and from quantum mechanical considerations it can be shown that the interaction energy E_{int} between the ith and jth electron spin in a given solid can be written as

$$E_{int} = -2\mathscr{J}S_i S_j \cos \phi_{ij}$$

where S_i and S_j are the spin values and ϕ_{ij} is the angle between the magnetic moments of the two spins. This formula is quite general, since intramolecular fields arise in substances other than ferromagnets, but it is only in the latter that the spins align in the same direction.

It is possible to determine the exchange integral $\mathscr{J}$ experimentally as a function of the ratio of the interatomic spacing a to the unfilled shell radius r, and this is shown for a number of atoms in Figure 3.14. It can be seen that $\mathscr{J}$ can take either positive or negative values. If $\mathscr{J}$ is positive, the material is ferromagnetic since the interaction energy has a minimum value when $\phi_{ij} = 0$ and $\cos \phi_{ij} = 1$. Most of the materials normally referred to as ferromagnetics, however, have not only positive but also large exchange integrals and are already well below their Curie points at room temperature (see below). On the other hand, an element such as gadolinium has a very low exchange integral and has a Curie temperature (289 K) at just about room temperature. Such an element therefore should be classed as ferromagnetic, but does not show such behaviour at room temperature.

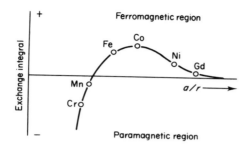

Figure 3.14. Variation of exchange integral with a/r

The state of ferromagnetism corresponds to minimum energy. On the other hand, when $\mathscr{J}$ is negative, the interaction energy is only a minimum when the angle between adjacent spins is 180° (that is when any two spins are oppositely directed to one another). This state is called *antiferromagnetism*. It is interesting to note that manganese can be made ferromagnetic by alloying to form compounds whose a/r ratios are greater than that of pure manganese.

The intramolecular field tending to align the spins within a domain is highly dependent on temperature. At low temperatures all the spins are pointing in the same direction, but as the temperature is raised thermal agitation tends more and more to disarrange them until eventually they become completely random and the domain structure breaks down. The specimen then ceases to be ferromagnetic at some critical temperature Θ_c called the *Curie temperature* (about 1000 K for iron) and behaves paramagnetically. The susceptibility then follows the Curie law, although for ferromagnetic materials this is termed the Curie–Weiss law:

$$\chi_m = \frac{C_c}{T - \Theta_c}$$

where C_c is the Curie constant. For temperatures above Θ_c the material ceases to be ferromagnetic and becomes paramagnetic as shown in Figure 3.15.

Movement of domain walls

The nature of the boundaries separating the domains will now be considered more closely. As has been pointed out, the boundaries are regions in which the spins change their spatial directions. The transition layer between two domains is called a *Bloch* wall, and in iron this covers a zone of about 300 lattice spacings.

Figure 3.16 shows the structure of a Bloch wall for a transition from a domain with a spin-up to one with a spin-down arrangement. If a cross-section were taken through this specimen, the domain pattern would appear as shown in Figure 3.17(a). It should be realized that the Bloch walls are not rigidly fixed,

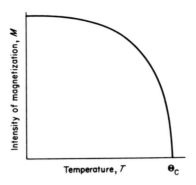

Figure 3.15. Variation of the intensity of magnetization with temperature for a ferromagnetic material

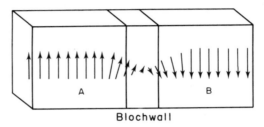

Blochwall

Figure 3.16. Schematic representation of a Bloch wall

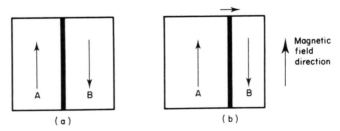

Figure 3.17. Motion of the Bloch wall resulting from the application of a magnetic field

but can move through the body of the sample. This means that one domain may increase in volume whereas another may decrease, as illustrated in Figure 3.17(b). One method of causing domain boundaries to move is to apply an external magnetic field to the specimen.

A completely demagnetized ferromagnetic specimen, represented schematically by the four equal domains of Figure 3.18(a), has no resultant magnetic moment.

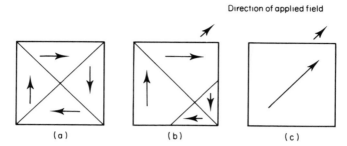

Figure 3.18. Magnetization by grain growth followed by spin rotation

If a gradually increasing field is applied in some arbitrary direction, as shown in Figure 3.18(b), those domains that have a magnetization component in the direction of the applied field will grow at the expense of the remaining two. At some stage, however, the field intensity will become sufficiently high to cause the resultant magnetic moments to rotate within the two domains, and the specimen will just consist of one large domain whose moment points in the direction of the field (Figure 3.18c). This would be the maximum achievable magnetization with this specimen, and no further increase in field could cause **M** to become larger. In this state, the specimen is said to be saturated.

This motion of Bloch walls explains qualitatively at least the shape of the **M–H** curve (Figure 3.12) mentioned at the beginning of this section. Such a curve has three distinct regions, as indicated in the schematic diagram of Figure 3.19:

(a) A region in which the motion of the Bloch walls is reversible, so if the field is decreased the fall in magnetization follows the same curve (a).
(b) A region in which the Bloch wall motion is irreversible, so if the field is decreased the magnetization does not follow the same curve (b).
(c) A rotational region leading to saturation, in which the domain boundaries stop moving and the individual magnetic moments of the domains come into alignment with the applied field direction (point c).

The reason for the irreversible domain boundary motion is that no given sample will be free from lattice defects and, particularly, the strain fields caused by such defects. It has been found that these strain fields have the effect of 'holding back' the Bloch walls, so a certain amount of energy is required to overcome these 'obstacles' to domain boundary motion. Once these obstacles are surmounted, any motion in the opposite direction by the boundaries is likely to be blocked. This irreversible motion continues until the rotation region is reached, after which saturation sets in.

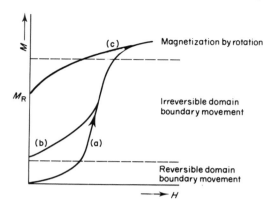

Figure 3.19. Hysteresis curve explained in terms of domain boundary movement

The hysteresis exhibited by ferromagnets is thus easily explained in terms of the strain field. When the domains start to move, a greater field in the reverse direction must be applied to cause the Bloch wall to surmount obstacles. If, for example, the field is completely removed, the intensity of magnetization returns to a value called the residual magnetization M_R; this means that not all the domain walls have returned to their original positions and some large volumes in the specimen remain which are magnetized in the direction of the applied field.

The 'locking-in' effect of these strain fields is illustrated by the experimental fact that a specimen that has been subject to severe mechanical treatment, such as hammering, can become a good permanent magnet. The mechanical working sets up more strains in the volume of the sample and these impede boundary motion.

3.5.4 FERRIMAGNETISM

A disadvantage of ferromagnetic materials in alternating current applications is their relatively low electrical resistivity, which leads to eddy current losses. For example, dynamo iron laminations have a resistivity of about 14×10^{-4} Ω m, and the highest attainable resistivity in a ferromagnetic alloy is about $10^{-2} \Omega$ m. On the other hand, ferrimagnets combine a permanent magnetic moment with a resistivity that can lie in the range $10-10^7 \Omega$ m.

The supreme example of ferrimagnetic materials is that class of oxides which contain trivalent ions with the structure of the mineral spinel ($MgAl_2O_4$). The term ferrospinel, or ferrite, is applied to the group of iron oxides that have the general formula $MO.Fe_2O_3$ where M is a divalent metal such as Mn^{2+}, Fe^{2+}, Co^{2+}, Ni^{2+}, Cu^{2+}, Mg^{2+} or Cd^{2+}. The typical ferrite is magnetite (Fe_3O_4 or $FeO.Fe_2O_3$), which has been known to have magnetic properties since ancient

Table 3.6. Magnetic moments (in Bohr magnetons) for ferrites with the inverse spinel structure

Ferrite	A site			B sites						Total moment of molecule
	Ion	m	Direction	Ion	m	Direction	Ion	m	Direction	
Manganese	Fe^{3+}	5	→	Mn^{2+}	5	↑	Fe^{3+}	5	↑	5
Iron	Fe^{3+}	5	→	Fe^{2+}	4	↑	Fe^{3+}	5	↑	4
Cobalt	Fe^{3+}	5	→	Co^{2+}	3	↑	Fe^{3+}	5	↑	3
Nickel	Fe^{3+}	5	→	Ni^{2+}	2	↑	Fe^{3+}	5	↑	2
Copper	Fe^{3+}	5	→	Cu^{2+}	1	↑	Fe^{3+}	5	↑	1

times. If the divalent iron ion in Fe_3O_4 is replaced by another divalent ion, various ferrites can be produced that have different intensities of intrinsic magnetization. The chemical formula of the mineral spinel may be written as

$$MgAl_2O_4 = MgO.Al_2O_3 = Mg^{2+}(Al^{3+}Al^{3+})O_4^{2-}$$

The material forms ionic crystals of cubic structure in which the oxygen anions have a face centred cubic arrangement (see Section 2.5.1) accommodating the smaller positive ions in the interstices. Of the two available sites, the Mg^{2+} ions occupy the 'A sites', in which they have four near-neighbour oxygen ions, whilst the Al^{3+} ions fill the remaining 'B sites' with six near-neighbour oxygen ions. The spinel structure is shown in Figure 3.20. In ferrospinels of normal structure, Fe^{3+} ions take the place of the Al^{3+} ions and divalent ions of another element are substituted for Mg^{2+} (for example, $ZnFe_2O_4$ and $CdFe_2O_4$). A zinc ferrite might therefore be formulated as $Zn^{2+}(Fe^{3+}Fe^{3+})O_4^{2-}$.

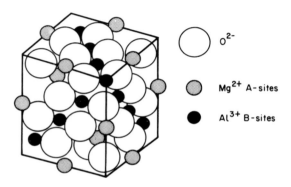

Figure 3.20. The spinel structure

The magnetic ferrites have a different siting of the cations and are said to have an inverse spinel structure. Again, there are two trivalent iron ions, one in a B site and the other in an A site. Also occupying a B site is a divalent ion of an element such as manganese, iron, cobalt or nickel. Denoting the latter ion by M^{2+}, the formula for an inverse structure ferrite can be written

$$MFe_2O_4 = MO.Fe_2O_3 = Fe^{3+}(M^{2+}Fe^{3+})O_4^{2-}$$

the brackets enclosing the ions on the B sites.

Three types of interactions are possible. It is considered that the interactions between A ions and A ions and between B ions and B ions are comparatively weak and that AB type reactions predominate. In fact, in the inverse spinel structure, the A and B sites are quite distant from one another and, in order to account for the observed effects, it has been assumed that the oxygen in an intermediate site somehow manages to transfer the interaction from the

A magnetic atom to the B magnetic atom. Such an interaction is called superexchange, but in any event the situation is such that the exchange integral is negative so that antiparallel pairing occurs. In this case, however, since the magnetic moments of the A and B ions are not equal, a net magnetic moment arises for each molecule and the magnitude of this moment depends on the difference of the individual moments.

Table 3.6 shows the resultant magnetic moments m per molecule for a number of ferrites with the inverse spinel structure. The oxygen ions are not included in this table since they have no magnetic moment, having been doubly ionized. The magnetic ferrite crystal has a domain structure, as do ferromagnetic metals, so a ferrite could be represented schematically as in Figure 3.21. In spite of the permanent magnetic moment of the domains, a ferrimagnetic substance is not usually spontaneously magnetized but exists rather in a demagnetized state owing to the random orientations of the domains. However, just as for ferromagnets, an applied field causes the domains to orientate in the field direction and the removal of the field leaves the material in a state of permanent magnetization.

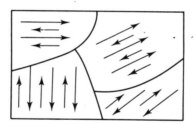

Figure 3.21. Domains in a ferrite

3.6 SUPERCONDUCTIVITY

To conclude this chapter, a few brief remarks will be made about superconductivity, a phenomenon discovered in 1911 by H. Kammerlingh-Onnes, which has far-reaching implications for an understanding of the electrical and magnetic processes in solids. In essence, the resistance of a superconductor decreases in a smooth fashion with temperature just like any normal metal. However, at some critical temperature T_c a phase transition occurs and the resistivity suddenly drops to zero, as shown in Figure 3.22, although an experimental upper limit is of the order of 10^{-25} Ω m. Moreover, it was further discovered by W. Meissner and R. Ochsenfeld in 1933 that an elemental superconductor always excludes all magnetic flux and is a perfect diamagnetic material. This of course shows that, although having zero resistance, a superconductor is not a *perfect* conductor. However, it does mean that once

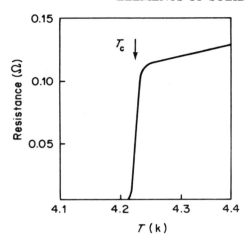

Figure 3.22. Typical variation of resistance with temperature for a superconductor

a current is established in, say, a ring of that material, it will continue virtually indefinitely without the need for a driving field.

The first material to show this effect was mercury, which becomes superconducting at liquid helium temperature, that is 4.2 K, and subsequently some 28 metals were found to be superconductors, with other elements becoming superconducting under special circumstances, such as structural disorder or when subjected to high pressures. In addition, the phenomenon has been demonstrated in thousands of intermetallic compounds and alloys, but until very recently the highest T_c achieved was of the order of 20 K.

What is interesting, however, is that it was not until 1957 that a theoretical basis for superconductivity was published by J. Bardeen, L. Cooper and R. Schreiffer (the BCS theory), for which they received the Nobel prize in 1972. Clearly the details are beyond the scope of this book and require a somewhat advanced appreciation of quantum mechanics, but in essence the key to the theory is the interaction between pairs of conduction electrons, called *Cooper pairs*. As was assumed in the Drude theory, a given electron will exert an attractive force on the positive ion cores, thereby slightly distorting the lattice. This slight concentration of positive charge then attracts another electron, which has a momentum opposite to that of the first. At ordinary temperatures the electron pair interaction is swamped by the normal thermal motion, but it can become significant at low temperatures, and since the electrons operate as a pair, an individual electron cannot gain or lose small amounts of energy, say, from an applied field or through collisions. They can therefore move through the lattice without any energy exchange in collisions, which of course is the origin of resistance.

The BCS theory essentially related to low-temperature superconductors, but in 1986, G. Bednorz and A. Müller observed superconductivity at 40 K in a cuprate layered structure of a Ba–La–Cu–O system, for which they too were awarded the Nobel prize, in fact, the most rapidly ever awarded Nobel prize for physics. The race was then on and by 1991 there had been over 20 000 publications relating to high-temperature superconductors, the then record being held by a ceramic material, a thallium-based copper oxide layered superconductor with a critical temperature of 125 K. The significance of the Bednorz and Müller discovery was not just a dramatic increase in T_c but also the discovery of what could be a completely new class of metals with quite different properties to normal metals, requiring modified BCS theory, or indeed a completely new theory. Even more puzzling was the fact that the highest temperature superconductors appear to be ceramic oxides — materials that are normally regarded as insulators.

It is expected that a room temperature superconductor will eventually be developed, implying enormous technological possibilities, for example loss-free long-distance power distribution systems and energy conversion devices, including generators and motors, faster computer switching, improved medical imaging, and frictionless transportation, with magnetically levitated railway carriages at last becoming a reality.

3.7 SUMMARY

We have seen in this chapter that Drude's notion that electrons are the agents of conduction was essentially correct, but that the details of his model were not. By applying the normal kinetic theory of gases, he successfully predicted values for the conductivity, confirmed Ohm's law and verified the Wiedmann–Franz law. The classical theory broke down, however, on the following points:

(a) The measured mean free paths were often about an order of magnitude greater than the expected values.
(b) Specific heat measurements indicated that only a small fraction of the electrons that were available took part in the conduction process.
(c) Hall effect measurements indicated that positive as well as negative charge carriers were apparently present in solids. In some materials the positive ones predominate; in others the negative.
(d) The conductive properties of semiconductors and insulators could not be accounted for in any way.
(e) The behaviour of superconductors could not be explained.

The removal of these difficulties did not occur until about a quarter of a century after Drude's initial work, when quantum mechanics had been developed and applied to the problem of conduction. Thus, another example can be added to those already described in Chapter 1 of the inadequacy of classical physics in relation to certain problems, and in Chapter 5 we shall examine how far quantum theory does go in overcoming these apparent obstacles.

PROBLEMS

3.1 The atomic weight of sodium is 23 and its density is $0.97 \times 10^3 \, \text{kg} \, \text{m}^{-3}$. Calculate the number of electrons in a cubic metre of sodium.

3.2 Given a current density limit of $10^8 \, \text{A} \, \text{m}^{-2}$ in a thin aluminium film, calculate the maximum current carried by an aluminium integrated circuit connection $1 \, \mu\text{m}$ thick and $10 \, \mu\text{m}$ wide.

3.3 Aluminium is trivalent with atomic weight 27 and density $2.7 \times 10^3 \, \text{kg} \, \text{m}^{-3}$, whilst the mean free time between electron collisions is $4 \times 10^{-14} \, \text{s}$. Calculate the current flowing through an aluminium wire 10 m long and $1 \, \text{mm}^2$ cross-sectional area when a potential of 2 V is applied to its ends.

3.4 Calculate the intercollision time at room temperature and drift velocity in a field of $100 \, \text{V} \, \text{m}^{-1}$ in sodium, whose conductivity is $2.17 \times 10^7 \, \Omega^{-1} \text{m}^{-1}$.

3.5 The resistivity of a certain material is $0.02 \, \Omega \, \text{m}$ whilst the Hall coefficient is $5 \times 10^{-4} \, \text{m}^3 \, \text{C}^{-1}$. Deduce all the information you can about this material, assuming a field of $1 \, \text{V} \, \text{m}^{-1}$.

3.6 The intercollision time in copper is $2.3 \times 10^{-14} \, \text{s}$. Calculate its thermal conductivity at 300 K.

3.7 In a Hall effect experiment on silver a potential of $13.5 \, \mu\text{V}$ is developed across a foil of thickness 0.05 mm when a current of 6.4 A is passed in a direction perpendicular to a magnetic field of 1.25 T. Calculate the Hall coefficient for silver, the electron density and the mobility.

3.8 In a cubic crystal it can be shown that the diffusion constant D is related to temperature by the expression

$$D = D_0 \, e^{-Q/kT}$$

where D_0 is a constant and Q the activation energy, that is the energy needed for an atom to make a jump.

The following data refer to the diffusion of zinc in copper:

T(K)	$D \, (\text{m}^2 \, \text{s}^{-1})$
1322	1.0×10^{-12}
1253	4.0×10^{-13}
1176	1.1×10^{-13}
1007	4.0×10^{-15}
878	1.6×10^{-16}

Determine the constant D_0 and activation energy Q.

3.9 Assuming that the mean free path in copper decreases with the impurity concentration, estimate the resistivity of copper containing 2% Ni. The mean free path for impurity collision in copper containing 1% Ni is 60 nm, the electron thermal velocity at 300 K is 1.2×10^5 m s^{-1} and the resistivity of pure copper is 1.72×10^{-8} Ω m.

4 Energy Bands in Solids

In the previous chapter, several of the shortcomings of the classical theory of conduction were highlighted and we saw in Chapter 1 that certain experiments could only be interpreted by supposing that electrons sometimes behaved as though they had wave-like properties. We shall now extend this concept and develop a non-rigorous mathematical description of electron waves.

4.1 THE WAVE EQUATION

Since de Broglie had suggested that electrons had a wave nature, Erwin Schrödinger in 1926 argued that it should be possible to represent them mathematically by a suitable wave equation. The concept of waves and the basic forms of the wave equations are well known from other branches of physics, for example sound and light waves, and it is assumed that the reader is familiar with the basic mathematical relationship for a one-dimensional travelling wave.

Let the quantity that varies in the 'electron wave' be Ψ (pronounced 'psi') which is a function of both position and time and whose significance will be discussed later. Then, confining the motion to the x direction, we may write

$$\Psi = \Psi_0 \sin 2\pi\left(\frac{x}{\lambda} - \frac{t}{T}\right) \tag{4.1}$$

where the constants Ψ_0, λ and T are the amplitude, wavelength and period of the motion respectively. The period $T = 1/\nu$, where ν is the frequency.

If equation (4.1) refers to an electron or beam of electrons which can be represented by a de Broglie wave, de Broglie's and Planck's equations ($p = h/\lambda$ and $E = h\nu$) can be used to rewrite it as

$$\Psi = \Psi_0 \sin 2\pi\left(\frac{xp}{h} - \frac{Et}{h}\right) \tag{4.2}$$

Alternatively, a more general expression for a wave takes the form

$$\Psi = \Psi_0 \exp i(xp - Et)/\hbar \tag{4.3}$$

where $i = \sqrt{-1}$ and $\hbar = h/2\pi$. Differentiating equation (4.3) twice with respect to x gives

$$\frac{\partial^2 \Psi}{\partial x^2} = -\Psi \frac{p^2}{\hbar^2} \tag{4.4}$$

117

and, differentiating once with respect to t, gives

$$\frac{i\hbar\partial\Psi}{\partial t} = E\Psi \tag{4.5}$$

The kinetic energy of a particle with momentum p and mass m moving slowly compared with the velocity of light is $p^2/2m$, so that if E is the total energy of the particle and V its potential energy we have

$$\frac{p^2}{2m} + V = E \tag{4.6}$$

Multiplying each term in equation (4.6) by Ψ and substituting for p^2 and E from equations (4.4) and (4.5) gives

$$-\frac{\partial^2\Psi}{\partial x^2}\left(\frac{\hbar^2}{2m}\right) + V\Psi = i\hbar\,\frac{\partial\Psi}{\partial t} \tag{4.7}$$

This is known as the 'time-dependent' Schrödinger wave equation (for one dimension). The use of this equation, rather than Newton's second law, to solve a dynamical problem is similar to the use of wave optics rather than geometrical optics in solving a problem in light. It should be stressed that the above is not a derivation but rather a justification of the Schrödinger equation.

4.1.1 STATIONARY STATES

When using the Schrödinger wave equation on an atomic problem, the principle is to substitute the appropriate function for V and solve for Ψ, taking into account any special conditions which might exist (known as boundary conditions). The resulting solutions for Ψ are known as *wave functions*, and the energies associated with them are called *eigenvalues*.

Let us consider a possible wave function $\Psi(x,t)$ which can be expressed as a product of functions in x and t separately; for example,

$$\Psi(x,\ t) = \psi(x)\chi(t)$$

Substituting this expression into equation (4.7) and rearranging, we obtain

$$-\frac{\hbar^2}{2m\psi}\left(\frac{d^2\psi}{dx^2}\right) + V = \frac{i\hbar}{\chi}\frac{d\psi}{dt} \tag{4.8}$$

If V is a function of x only (that is if V is independent of time), the left-hand and right-hand sides of equation (4.8) are dependent on only x and t respectively, and so both sides must equal the same constant, let us say K. Thus

$$\frac{i\hbar}{\chi}\frac{d\chi}{dt} = K \tag{4.9}$$

which on integration gives us

$$\chi = \chi_0 \exp\left(\frac{-iKt}{h}\right) \tag{4.10}$$

where χ_0 is a constant of integration which is normally put equal to unity. If we compare this equation with equation (4.3) (for which Ψ is separable) we can see that K is identifiable with E, the total energy. Under these conditions, Ψ is said to represent a *stationary state* of the system. If we put $K = E$ and substitute equation (4.9) into equation (4.8), we obtain

$$\frac{d^2\psi}{dx^2} + \left(\frac{2m}{h^2}\right)(E - V)\psi = 0 \tag{4.11}$$

This expression for ψ is known as the 'time-independent' Schrödinger equation (in one dimension). Solutions for ψ are also known as wave functions, and the values of associated energies E, as we have seen, are known as eigenvalues.

A simple analogy is useful in appreciating the difference between Ψ and ψ. Consider a string of length L rigidly held at both ends. These boundary conditions cause stationary waves with specific wavelengths to be set up such that $n\lambda/2 = L$, where n is an integer. Such a standing wave might be represented by

$$\Psi = \Psi_0 \sin\left(\frac{2\pi x}{\lambda}\right)\cos\left(\frac{2\pi t}{T}\right)$$

or

$$\Psi = \Psi_0 \sin\left(\frac{2\pi xn}{2L}\right)\cos\left(\frac{2\pi t}{T}\right) \tag{4.12}$$

where Ψ represents the transverse displacement of the string. Here Ψ is the product of the functions $\psi(x)$ and $\chi(t)$, where

$$\psi(x) = \Psi_0 \sin\left(\frac{2\pi xn}{2L}\right) \quad \text{and} \quad \chi(t) = \cos\left(\frac{2\pi t}{T}\right)$$

Each point on the string vibrates in simple harmonic motion, and the amplitude $\psi(x)$ of this motion, although varying from point to point along the string, remains constant in time. The form of equation (4.12) shows that only certain wave functions $\psi(x)$ (and therefore motional energies) are possible. This point will be discussed in greater detail below.

4.1.2   PHYSICAL INTERPRETATION OF Ψ

The symbol Ψ represents a complex quantity with real and imaginary parts. It is related to the requirement that it must completely represent the motion, and must specify where the electron is and what energy it has at a given time.

The interpretation given by M. Born (1920) is that $|\Psi|^2 dx$ represents the probability of finding an electron between a distance x and $x + dx$ from the origin. However, since the electron also carries a charge, the function also represents the charge distribution associated with that electron, and this is quite a helpful concept. Where the charge density is greatest is the most likely place to find the electron.

In three dimensions the Schrödinger equation becomes

$$\frac{\partial^2 \Psi(x,y,z)}{\partial x^2} + \frac{\partial^2 \Psi(x,y,z)}{\partial y^2} + \frac{\partial^2 \Psi(x,y,z)}{\partial z^2} + [E - v(x,y,z)]\, \Psi(x,y,z) = 0 \qquad (4.13)$$

and

$$|\Psi|^2 \, dx \, dy \, dz$$

is the probability of finding the electron in the volume $dx\, dy\, dz$ at a given instant. Moreover, since the electron must be somewhere, the integral of the above probability expression over the appropriate region of space where the electron might exist must be unity; that is

$$\int |\Psi|^2 \, dx \, dy \, dz = 1 \qquad (4.14)$$

It is usual to use $|\Psi|^2$ rather than Ψ for the probability density, to allow for the fact that the wave function could be a complex quantity.

The idea that an electron had a certain probability of existing at any given distance from, say, the proton in the hydrogen atom was difficult to comprehend. In terms of the Rutherford atom the electron was a solid particle at a fixed distance from the nucleus. A plot of the probability of finding the electron for that model would be as shown in Figure 4.1(a), that is zero probability until the particle is encountered and zero after it has been encountered. Figure 4.1(b) illustrates the analogous case for the wave representation.

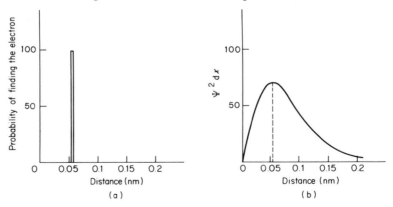

Figure 4.1. Probability of finding an electron. (a) Particle nature, (b) wave nature

Near zero radius there is little chance of finding the electron, but at 0.053 nm $|\Psi|^2 dx$ reaches its maximum (that is at the first Bohr radius), and then decreases asymtotically at larger distances. However, at any distance, no matter how near or far from the proton, there is always a finite chance that the electron will be there.

Clearly the implication that it is never possible to be absolutely certain where an electron is at a given instant is one manifestation of the uncertainty principle, introduced in Chapter 1. Moreover, a real particle that is highly localized cannot be described by a wave of the type given by equation (4.1). It is more appropriate to build up a wave packet using individual waves of carefully selected amplitudes and frequencies which then destructively interfere except for that region of space where the probability of finding the particle is high. It will be recalled from equation (1.14) that the speed of the particle will be the same as the group velocity of the wave packet

$$v_g = \frac{\partial \omega}{\partial k}$$

where ω is the angular frequency of the de Broglie waves and k is the wave number or vector. Figure 1.8 showed such a wave packet localized in a region of space dx. The probability of finding the particle will be a maximum at the centre of the packet, although there is a small but finite chance of finding it anywhere in the region.

4.2 SOLUTION OF THE SCHRÖDINGER EQUATION FOR A PARTICLE IN A ONE-DIMENSIONAL BOX

In the hydrogen atom, the electron is tightly bound — or in other words is trapped by the nucleus. A very crude model can be set up for the hydrogen atom on the supposition that a particle of mass m (representing the electron) moves in an infinitely deep potential well. In terms of a particle, this would correspond to, say, a ball at the bottom of an infinitely tall tube. If the tube were shaken around, various energies could be imparted to the ball but it nevertheless would remain trapped. Before considering this three-dimensional problem, the Schrödinger equation will be applied to the one-dimensional case and the existence of energy levels for a bound particle will be predicted.

The essential task in solving the equation is to establish an appropriate potential function $V(x)$, try various periodic solutions $\psi(x)$ until one that satisfies the equation is found and then calculate the associated energy eigenvalues E.

The potential function shown in Figure 4.2 can be defined as follows:

$$V(x) = \infty \qquad \text{for } x < 0 \text{ or } x > L$$

and

$$V(x) = 0 \qquad \text{for } 0 < x < L$$

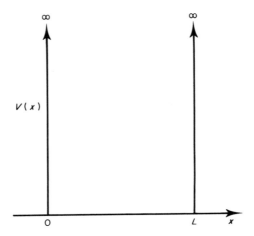

Figure 4.2. The one-dimensional potential box

The choice of $V(x) = 0$ between 0 and L is quite arbitrary and only serves as a constant reference level in the same way as sea level is taken as an arbitrary datum from which terrestrial heights can be measured. In this restricted interval $0 < x < L$, the wave equation (equation 4.13) becomes

$$\frac{d^2\psi(x)}{dx^2} + \frac{8\pi^2 m}{h^2} E\psi(x) = 0$$

For the sake of simplicity a sine function can be chosen as an appropriate periodic function. Thus

$$\psi(x) = A \sin(kx)$$

will be a satisfactory solution, provided that $k = n\pi/L$, where n is an integer, since it reduces the wave function to zero at both the boundaries. The condition that solutions to Schrödinger's equation should exist is thus

$$k = \frac{n\pi}{L}$$

so that

$$\psi(x) = A \sin \frac{n\pi x}{L} \tag{4.15}$$

where $n = 1,2,3, \ldots, \infty$. Substitution of this solution into equation (4.14) gives

$$E = \frac{h^2 k^2}{8\pi^2 m} = \frac{h^2 n^2}{8mL^2} \tag{4.16}$$

Therefore, since k is restricted to certain values, the total energy E must also have a similar restriction.

The parameter n which determines the energy in equation (4.16) is equivalent to the quantum number met in the Bohr theory of the hydrogen atom in Chapter 1. If n is allowed to take various values, solutions very similar to those for a vibrating string are obtained; instead of the continuous range of solutions found for a free particle, standing waves result, each associated with a specific energy value or level. The first three solutions to the problem of electron waves are illustrated in Figure 4.3, together with the electron probability distributions $|\psi|^2$ and energy levels.

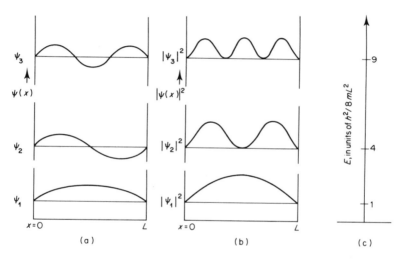

Figure 4.3. Electron in a box, ground and first two excited states: (a) electron wave functions; (b) electron probability distributions; (c) electron energy levels

It should be noted that these distributions are a two-dimensional way of representing a one-dimensional charge distribution. The plot for the lowest energy tells us that the charge density is a maximum in the centre of the box and falls off towards the ends of the box. Just as for a vibrating string, the more loops there are, the higher the associated energy. On this basis, a simple energy level diagram can be constructed for a particle in a one-dimensional box (Figure 4.3c). The level with the lowest energy (that is the single loop) will represent the ground state, and all higher energies will correspond to excited states.

A trapped or bound particle has, therefore, a discrete energy spectrum. If the box length L is made infinitely large, the particle is essentially free and the energy differences between levels become smaller and smaller so that the energy spectrum becomes continuous. Thus, the wave equation predicts that bound

particles are associated with a discrete energy spectrum and free particles with
a continuous spectrum.

4.3 SOLUTION OF THE SCHRÖDINGER EQUATION FOR A PARTICLE IN A THREE-DIMENSIONAL BOX

The one-dimensional box so far considered is clearly unsatisfactory as a model
for the hydrogen atom, which is three dimensional. It would therefore seem
reasonable to extend the potential box to the y and z directions. The equation
to be solved then becomes

$$\frac{\partial^2 \psi(x,y,z)}{\partial x^2} + \frac{\partial^2 \psi(x,y,z)}{\partial y^2} + \frac{\partial^2 \psi(x,y,z)}{\partial z^2} + \frac{8\pi^2 m}{h^2} E\psi(x,y,z) = 0 \qquad (4.17)$$

Although apparently more complex than equation (4.13), this yields
comparatively straightforward solutions, again of the standing wave type:

$$\psi(x,y,z) = A \ \sin(k_1 x)\sin(k_2 y)\sin(k_3 z)$$

where $k_1 = n_1\pi/L$, $k_2 = n_2\pi/L$, $k_3 = n_3\pi/L$. Each number n_1, n_2, and n_3 can take
any integral value from one to infinity. The energy term is therefore governed
by three quantum numbers so that equation (4.16) becomes

$$E = \frac{h^2}{8mL^2}(n_1^2 + n_2^2 + n_3^2)$$

$$= \frac{n^2 h^2}{8mL^2} \qquad (4.18)$$

where

$$n^2 = n_1^2 + n_2^2 + n_3^2$$

These quantum numbers also determine the form of the wave function which,
by convention, can be written $\psi_{n_1 n_2 n_3}$. The most important consequence of
these three quantum numbers is that several combinations can yield the same
value of energy. For example, suppose one of the quantum numbers is equal
to two, the others being unity. Three possible combinations of the quantum
numbers lead to the same value of energy E, namely:

(a) $n_1 = 1$, $n_2 = 1$ and $n_3 = 2$;

(b) $n_1 = 1$, $n_2 = 2$ and $n_3 = 1$;

(c) $n_1 = 2$, $n_2 = 1$ and $n_3 = 1$.

The corresponding wave functions are then

$$\psi_{112} = A\left(\sin\frac{\pi x}{L}\right)\left(\sin\frac{\pi y}{L}\right)\left(\sin\frac{2\pi z}{L}\right)$$

$$\psi_{121} = A\left(\sin\frac{\pi x}{L}\right)\left(\sin\frac{2\pi y}{L}\right)\left(\sin\frac{\pi z}{L}\right)$$

and

$$\psi_{211} = A\left(\sin\frac{2\pi x}{L}\right)\left(\sin\frac{\pi y}{L}\right)\left(\sin\frac{\pi z}{L}\right)$$

but in each instance the energy is exactly the same, that is

$$E = \frac{6h^2}{8mL^2}$$

If several states have the same energy, they are said to be *degenerate*. In the above example, since three wave functions are associated with the same energy, the corresponding energy level is said to be *threefold* degenerate. On the basis of this model, any level in which the three quantum numbers are equal would be non-degenerate.

Figure 4.4 shows the energy level diagram for a particle in a three-dimensional cubic box for the ground state and some excited states, together with the degrees of degeneracy and quantum numbers.

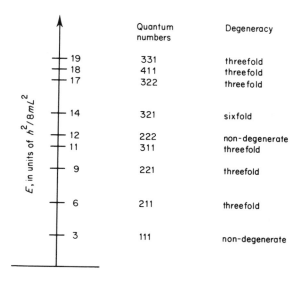

Figure 4.4. Energy levels for an electron in a three-dimensional box

Example 4.1

Calculate the amount of energy emitted when an electron in a three-dimensional box makes a transition from the first excited state to the ground state. Assume the box is cubic of side length 0.5 nm.

The eigenvalues for a three-dimensional box are given by equation (4.18) as

$$E = (n_1^2 + n_2^2 + n_3^2)\frac{h^2}{8mL^2}$$

For the ground state the quantum numbers are (111), while for the first excited state they are (112). Hence the energy emitted is

$$E_1 - E_g = 3(6.6 \times 10^{-34})^2/8 \times (9.1 \times 10^{-31}) \times (0.5 \times 10^{-9})^2$$
$$= 7.18 \times 10^{-19} \text{ J} = 4.5 \text{ eV}$$

In practice, the three-dimensional box is only an approximation to the hydrogen atom, which has spherical rather than cubic symmetry. However, the Schrödinger equation can be transformed quite readily to spherical polar coordinates and solved in a manner analogous to that for the cubic box.

The argument can be extended even further by solving the time-dependent equation, and it is found that a fourth quantum number is needed to specify the wave function. However, the allowed values of these numbers are the same as those already introduced in Chapter 1 and referred to in Chapter 3, namely n, l, m_l and m_s.

Thus, without making arbitrary assumptions, as was the case in the older quantum theory, wave mechanics has allowed us to predict and calculate the energy levels in atoms. However, unlike the one-dimensional case, it is more difficult to visualize the charge density distribution for three dimensions. Clearly, the plots will represent surfaces up to some specific value of charge density, for example, for the case of the ground state of the hydrogen atom, to the first Bohr radius (Figure 4.1b). Figure 4.5 shows the angular dependence of the probability distribution functions at some arbitrary distance from the centre of the hydrogen atom. The ground state, the so-called s orbital, with quantum numbers $n = 1$, $l = 0$, $m_l = 0$ is clearly spherically symmetric, whereas the first excited state with $n = 2$, $l = 1$ and $m_l = \pm 1$, 0 has three equivalent distributions for each coordinate direction, the so-called p orbitals.

The principal predictions resulting from the wave mechanical treatment of electrons bound in atoms can thus be summarized as follows:

(a) Electrons can only exist in particular specific states, termed energy levels, corresponding to specific energy values, each governed by a certain set of quantum numbers.

(b) It is possible for a given energy level to be degenerate (that is more than one electron probability distribution may be associated with the same energy).

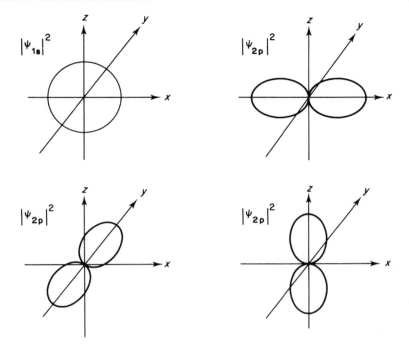

Figure 4.5. Electron probability distributions $|\psi_{1s}|^2$ and $|\psi_{2p}|^2$ for the ground and excited states of the hydrogen atom

(c) The actual distribution of electron charge in space is found by plotting the square of the amplitude of the wave function in three dimensions.

4.4 REMOVAL OF DEGENERACY

As we saw in the discussion of the three-dimensional box problem, most energy levels are degenerate unless they have three identical quantum numbers. In fact, the introduction of the fourth quantum number m_s means that every level must be at least twofold degenerate, since m_s can have the two values $\pm\frac{1}{2}$. All atomic energy levels therefore consist of two or more overlapping energy states, and some of the higher levels (with large values of n) must be highly degenerate.

However, it is possible to 'remove' the degeneracy of an energy level by splitting it into a set of closely separated sublevels. This can be achieved by applying some external agency to the atom, such as a magnetic or an electric field, which has the effect of depressing some and raising other degenerate levels within a given electronic state. A well-known example in which the spectral lines emitted from a light source subjected to a strong (≈ 1 T) magnetic field are split

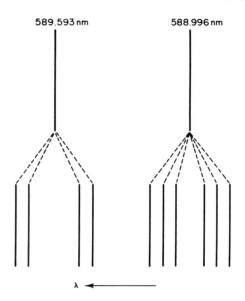

Figure 4.6. Diagram illustrating the Zeeman effect for the sodium D spectral lines

into a number of components was discovered in 1896 by P. Zeeman. Sodium light provides a good example of Zeeman splitting, as is illustrated in Figure 4.6. The splitting in energy levels of any given atom giving rise to these lines is illustrated in Figure 4.7, from which it can be seen that the two originally degenerate 3P levels have split into four components and two components, and the twofold degenerate 3S levels into two components. It should be noted that the designation of S and P levels is a nomenclature taken from the earlier theories of atomic energy levels, and that these diagrams are meant to illustrate the principles of the process, rather than the details.

An analogous effect when a strong electric field was applied was observed by J. Stark in 1913.

The removal of degeneracy as a result of the application of external agencies is the key to the reason why energy bands occur in solids.

4.5   AGGREGATES OF ATOMS

4.5.1   ORIGIN OF ENERGY BANDS

We shall now consider the changes occurring in electronic energy levels when a group of identical atoms with initially wide physical separations are gradually brought closer and closer together to form a solid, that is when condensation from the gaseous to the solid phase takes place.

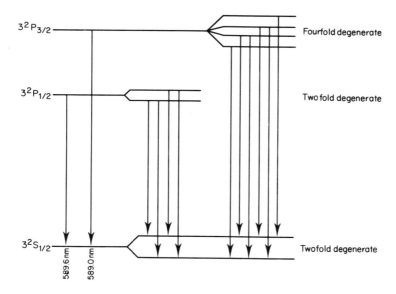

Figure 4.7. Energy level diagram for sodium showing the origin of the D lines and the level splitting which occurs in a magnetic field

At the outset the energy levels of any given atom are identical to those in the free state, but as the spacing is gradually reduced a stage is reached at which the outer valence electrons begin to interact with the valence electrons in other atoms. The external fields produced by the electrons of neighbouring atoms cause the valence levels to split in a manner similar to that observed in the Zeeman or Stark effects. As the interatomic spacing is still further reduced, the inner electronic levels also start to split. An energy level diagram as a function of separation is then as shown in Figure 4.8. The energy passes through a minimum at some specific interatomic spacing r_0 corresponding to the equilibrium position. This is to be expected, since all physical systems tend to a minimum energy state corresponding to stable equilibrium; this was discussed in Section 2.6 (Figure 2.23). It is reasonable that the valence level should begin to split at a larger separation than all the others because these electrons, being furthest away from the parent nucleus, are less tightly bound and more likely to be affected by external influences.

We can therefore see that the result of bringing atoms together to form a solid is that the discrete energy levels spread out into a range of sublevels which are so close together that they are termed an energy band. The band into which the ground state valence level splits is termed the valence band, and the lower lying bands corresponding to the inner lying electronic levels in the free atom are designated by the appropriate spectroscopic symbols. One therefore speaks

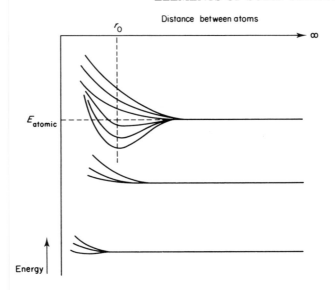

Figure 4.8. Diagram showing the splitting of energy levels due to the interaction of adjacent atoms as a function of interatomic separation; r_0 corresponds to the equilibrium interatomic spacing in the solid

of s bands, p bands and so on, but in the construction of energy band schemes of solids it is customary to omit the inner bands since they have little effect on the main electronic properties of the solid (see Section 5.2). This is entirely analogous to the free atom, in which the valence electrons are of primary importance.

The band corresponding to the level immediately above the free atom valence level (that is the first excited state) is called the conduction band and, just as for the excited state of a free atom, it may or may not be occupied at any given time. The question then arises as to how many energy levels exist within a given band? In the Zeeman effect, the external field perturbation in fact did not change the number of levels—only their separation. Thus the sodium 3S level, which originally consisted of two superimposed levels, was split from zero separation to a finite value. The total number of levels in a band must therefore be the number M of levels in an individual atom multiplied by the total number N of atoms in the solid (that is MN). Thus an s band will have $2N$ levels, a p band $6N$ levels, a d band $10N$ levels and so on.

A rough estimate of the average level separation within a band may be obtained as follows. It can easily be demonstrated that a crystal weighing 1 kg contains approximately 10^{26} atoms and, as will be shown below, bands are typically 1.0 eV wide. There are thus $M \times 10^{26}$ levels spread over 1.0 eV, so that the spacing is less than 10^{-26} eV. The energy levels within a band are therefore discrete, but of such small separation that the band can be taken as continuous to the first approximation.

The question now arises of how the electrons fit into the levels available within a band. As with isolated atoms, the Pauli principle, which states that no two electrons can have the same set of four quantum numbers, can be applied. In a band, the energy states are no longer coincident, but discrete and separate levels are available and one electron can be assigned to each of them. The lowest level is filled first, then the next highest and so on until all the available electrons are used. The Pauli principle therefore is still obeyed to within a band.

Considering the specific example of the sodium valence band: if the sodium crystal contains N atoms, the number of s subshell valence electrons will be N. In the band, however, $2N$ levels are available, so the valence band of sodium is only half filled. This important fact is discussed below in greater detail.

The band theory of solids therefore predicts the existence of a series of energy bands of finite width, E_b, separated from one another by certain energy gaps. Of major importance is the interval between the top of the valence and bottom of the conduction bands, which is called the forbidden energy gap and denoted by E_g. Neglecting the variation in band width and the lower lying states, the correlation between the free atom and the atom in a solid can be represented as in Figure 4.9. This topic is considered in greater detail in Section 5.1.

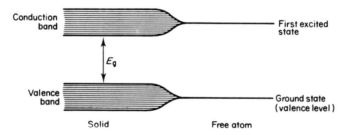

Figure 4.9. Correlation between energy levels for the free atom and energy bands for the solid

4.5.2   THE KRONIG–PENNEY MODEL

The previous section demonstrated in a qualitative way why energy bands form when atoms are aggregated to form a solid. However, it is also possible to predict the existence of bands by solving the Schrödinger equation for the periodic potential provided by a crystal lattice. A rigorous mathematical treatment is beyond the scope of this book, but the relatively simple one-dimensional model proposed by R. de L. Kronig and W. G. Penney in 1930 can be used to illustrate the origin of energy bands in a more quantitative way. The mathematical details will be found in Appendix 3, but essentially the model consists in solving the

Schrödinger equation for a periodic one-dimensional lattice consisting of a series of rectangular potential barriers of width b, height V_0 and separated by a distance of a, the lattice constant. The crucial consideration is what is the 'barrier strength'.

The equation that expresses the conditions for solutions to exist, subject to the chosen potential and boundary conditions, is

$$\frac{P \sin \alpha a}{\alpha a} + \cos \alpha a = \cos ka \qquad (4.19)$$

where $\alpha^2 = 8\pi^2 mE/h^2$ and k is the wave number. The quantity P is defined by

$$P = \frac{4\pi^2 ma(V_0 b)}{h^2}$$

and is a measure of the barrier strength, that is it depends on the barrier 'area', $V_0 b$. This means that if P is increased the binding of a given electron to a particular potential well is also increased.

Equation (4.19) therefore gives the values of α which permit solutions of the electron wave to exist. Since α is a function of energy, this means that the electron energies are restricted to certain values, which depend on the barrier strength. To illustrate, it is customary to choose $P = 3\pi/2$ and then Figure 4.10 shows a plot of the left-hand side of the equation against αa. Since the right-hand side of the equation can only have values between $+1$ and -1, the only accepted values of αa are those shown by the heavy horizontal lines in the diagram. Moreover, since αa is proportional to energy, the model predicts that the motion of an electron in a periodic lattice is characterized by 'bands' of allowed energies separated by forbidden regions. It should also be noted that as P increases, the energy band widths become narrower and ultimately reduce to the energy levels of a particle in a box of atomic dimensions. On the other hand, when P decreases

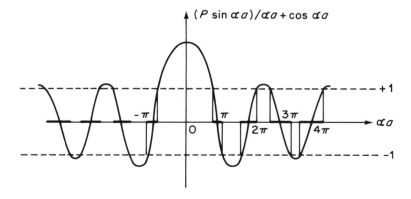

Figure 4.10. Plot of $(P \sin \alpha a)/\alpha a + \cos \alpha a$ against αa for $P = 3\pi/2$

to zero, $\alpha = k$, and the energy spectrum becomes that of a free electron, that is continuous, with

$$E = \frac{h^2 k^2}{8\pi^2 m}$$

The Kronig–Penney model therefore predicts that discontinuities will occur on the otherwise parabolic E–k curve when

$$k = \frac{n\pi}{a} \qquad \text{where } n = 1, 2, 3 \ldots,$$

as shown in Figure 4.11. These k values define the boundaries of what are termed *Brillouin zones*, intervals containing the values of the wave vector for individual energy bands. For example, the first zone contains k values lying between $-\pi/a$ and $+\pi/a$; the second the two segments π/a to $2\pi/a$ and $-\pi/a$ to $-2\pi/a$, and so forth. A simple interpretation is that since $k = 2\pi/\lambda$ the expression $ka = n\pi/a$ can be rewritten as $n\lambda = 2a$, which represents Bragg reflection for normal incidence of the electron waves on a set of planes of separation a (see Section 2.2).

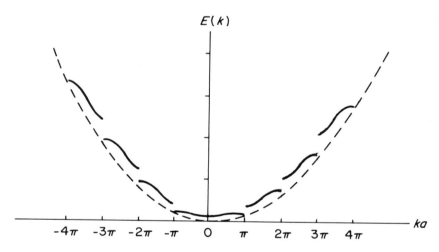

Figure 4.11. Plot of E against k for the Kronig–Penney model

Within an energy band, therefore, the energy is a periodic function of k. This means that if k is replaced by $k + 2\pi n/a$ where n is an integer, the right-hand side of equation (4.19) remains the same. For convenience, therefore, the E–k curve is often drawn in the reduced wave vector representation, which is limited to the region

$$-\pi/a < k < \pi/a$$

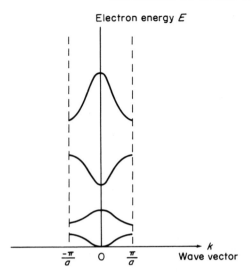

Figure 4.12. Reduced zone representation of the $E–k$ curve

This is shown in Figure 4.12. It should be noted that the situation is obviously much more complex in real three-dimensional solids, where the $E–k$ curves depend on the direction of the electron wave vector with respect to the crystallographic axes and the shape of the Brillouin zones depends on the crystal structure. They do, however, provide a very powerful tool for explaining the electronic and optoelectronic behaviour of semiconductors.

4.5.3   THE MOTION OF ELECTRONS

Although the motion of an electron through a crystal lattice as described by the Kronig–Penney model is an oversimplification of the actual potential found in real crystals, it can allow certain aspects of electron dynamics to be considered. For example, we have seen that the velocity of an electron wave packet with a specific k-value is given by

$$v_g = \frac{\partial \omega}{\partial k}$$

However, as the energy of the electron wave can be expressed as $E = \hbar\omega$, its velocity is given by

$$v_g = \hbar^{-1}\left(\frac{dE}{dk}\right)$$

which is the slope of the $E–k$ curve. Similarly, the acceleration, a, will be

$$a = \frac{dv}{dt} = \frac{1}{\hbar}\left(\frac{d^2E}{dk^2}\right)\left(\frac{dk}{dt}\right) \tag{4.20}$$

When an external electric field ε is applied for a period of time dt, an electron initially in a state specified by a given k value gains an amount of energy given by the work done on the electron, that is

$$dE = e\varepsilon v_g \ dt = \left(\frac{e}{\hbar}\right) \varepsilon \left(\frac{dE}{dk}\right) dt$$

Now $dE = (dE/dk)dk$ so the rate of change of k is given by

$$\frac{dk}{dt} = e\frac{\varepsilon}{\hbar}$$

The particle acceleration from equation (4.20) therefore becomes

$$a = \left(\frac{e\varepsilon}{\hbar^2}\right)\left(\frac{d^2E}{dk^2}\right)$$

Comparing this with the classical result for the acceleration of a free particle of mass, m,

$$a = \frac{e\varepsilon}{m}$$

it follows that the electron behaves as though it has an *effective mass* given by

$$m^* = \frac{\hbar^2}{d^2E/dk^2} \qquad (4.21)$$

This effective mass is therefore determined by the second derivative of the slope of the E–k curve, further illustrating the importance of such curves for the dynamics of electrons moving through crystal lattices. This is, of course, a mathematical model of effective mass, but equally well we could think of performing certain experiments in which the electron's dynamical behaviour could be explained by supposing that it has a mass different from the normal mass.

Figures 4.13(a), (b) and (c) show the energy, velocity and effective mass curves. An important consequence is that the effective mass is apparently positive in the upper part of the energy band and negative in the lower part, suggesting that above the point of inflection on the E–k curve the electron behaves like a positive particle. An interpretation of this will be considered in the next chapter.

4.6 THE FERMI LEVEL

In Section 4.5 it was explained how an energy band is filled and, for the specific example of sodium, it was shown that all the available levels within a band need not be occupied. This leads to a very useful model for describing energy bands.

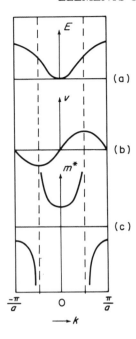

Figure 4.13. (a) Energy, (b) velocity and (c) effective mass as a function of the wave vector

An empty energy band can be likened to a stoppered flask whose height simulates energy and whose volume represents the number of available energy states (Figure 4.14). Another vessel contains water which represents the available number of electrons; these can be poured into the empty flask, thereby filling the available energy states, starting with the lowest. During this process, Pauli's principle is accounted for by the fact that no two drops can occupy the same

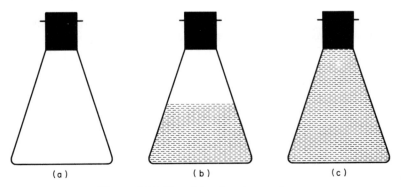

Figure 4.14. Fermi level water analogy

position. The volume of available water will clearly govern the position of the surface of the water and may result in a half-filled flask (band) or a completely filled flask (Figures 4.14b and c).

The energy corresponding to the surface level is called the Fermi level, and it is the boundary between filled and unfilled states. Since for the half-filled flask, even at room temperature, some drops near the surface will acquire sufficient energy to escape to the volume above (that is to make a transition to higher energy levels), the Fermi level is usually defined at absolute zero, when all thermal agitation ceases. The Fermi level is thus that level below which all states are occupied and above which all states are empty at absolute zero. It is important to realize, however, that the Fermi level need not always coincide with the top filled level, as will be seen for semiconductors.

Transition to higher states can be simulated on the basis of this same model by tilting or heating the flask; in terms of a real solid, these actions represent the application of electric fields or thermal energy. It is obvious that, in the situation represented by Figure 4.14(b), such transitions can easily be simulated, since a large volume is still available for occupation. On the other hand, in Figure 4.14(c), no amount of tilting or heating (unless very violent) can raise any drop to a higher state, since none is available for occupation. These two situations represent conductors and insulators respectively and are depicted in terms of energy level diagrams in Figure 4.15.

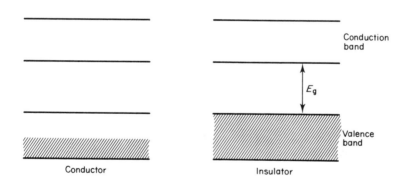

Figure 4.15. Diagram showing the energy band diagram of a conductor and an insulator (the shading represents occupied states in a band)

A knowledge of the Fermi level E_F, band width E_b, and energy gap E_g is of paramount importance in establishing the band picture of any given solid. A brief review of the experimental confirmation of the existence of bands and a discussion of some specific examples now follow.

4.7   EXPERIMENTAL INVESTIGATION
OF ENERGY BANDS

In order to study energy bands in solids, several modern sophisticated techniques can be used (such as cyclotron resonance, the de Hass van Alphen effect and magnetostriction), but the theory behind these methods is exceedingly complex and impossible to consider at this stage. However, there is one conceptually easy means of investigating bands, namely the use of soft X-rays.

X-rays are emitted when transitions take place between inner electronic levels. Before a transition can occur a vacant state must exist, and since the inner electrons are most tightly bound to the nucleus, a considerable amount of energy must be imparted to them to remove them from their stable orbits. This can be achieved by bombarding the atoms with high-energy electrons.

Consider the effect of a beam of high-energy electrons on an isolated atom, as shown in Figure 4.16. An inner electron is removed from an inner stable orbit—the 1s shell, say—leaving behind a vacant state. It is now possible for an electron in a higher state (for example the 2p state) to make the transition to the unoccupied level. In the process a photon of radiant energy is thereby emitted in a similar way to the emission of photons in the hydrogen atom when electrons make transitions between higher energy levels and the ground state of the valence electron (see Section 1.1.8). The energy associated with a transition is related to frequency by the expression $E_i - E_f = h\nu$ and, for the atom bombarded by electrons, this corresponds to an energy of about 80 eV which lies in the soft X-ray region.

A similar situation also arises when a solid is irradiated, except that the transition radiation then gives rise to a spectral band rather than a sharply defined spectral line.

Apparatus suitable for investigating the X-ray spectra of solids is shown in Figure 4.17. The electrons that are emitted by the source S strike the rotatable copper target T, on to which a thin film of the material to be studied can be

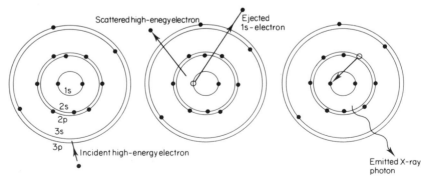

Figure 4.16. Emission of an X-ray photon from an atom (aluminium)

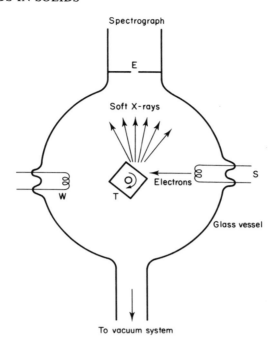

Figure 4.17. Apparatus for the investigation of soft X-ray spectra in solids

evaporated from a tungsten spiral W. The evaporation is carried out under vacuum, since the presence of oxides and other contaminants not only causes a deterioration of the surface but also alters the band structure. Even at a vacuum of 10^{-4} N m^{-2} (achieved by a conventional rotary-diffusion pump system plus liquid nitrogen traps to remove any excess gases evolved during evaporation) it is found that a surface soon becomes unsuitable for the investigations. However, by rotating the target, a fresh surface may be brought into coincidence with the X-ray beam. The radiation due to transitions in the material of the target passes through the entrance slit E of the spectrograph and is ultimately displayed as a spectrum on a photographic plate. Such a photograph can be interpreted in terms of the schematic energy level diagram for gaseous and solid sodium shown in Figure 4.18, in which only the most important levels are included.

In sodium, only the valence level is strongly affected when the atoms are put together to form a solid, the other inner levels remaining discrete. As a result, the sharp line arising from the 3s→2p transition in the gaseous state becomes a band of about 3 nm breadth in the solid state since transitions may take place from any level in the 3s band (the valence band) to the sharply defined 2p state.

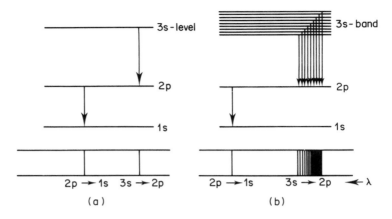

Figure 4.18. Schematic energy level schemes and X-ray spectra for: (a) gaseous and (b) solid sodium

From measurement of the breadth of the 3s→2p band and the sharp high-energy edge, the band width E_b and the Fermi level E_F with respect to the 2p level can be estimated. Some experimental results obtained in this way are shown in Table 4.1.

One of the most important results to emerge from the study of the X-ray spectra of solids is the distribution of electron states within the band itself. A soft X-ray spectrum photograph clearly shows that the distribution of states within a band is not uniformly dense across its entire width, but has a gradual increase in density towards the high-energy side. If such a photograph is scanned with a microdensitometer—an instrument that can measure the degree of 'blackening' on a photographic plate—curves similar to those of Figure 4.19 are obtained. Since an X-ray photon is emitted whenever an electron makes a transition from one energy level to another, the intensity of radiation at that wavelength must be proportional to the number of electrons making the

Table 4.1. Fermi levels measured from the bottom of the valence band

Element	Fermi level (eV)
Lithium	3.7
Sodium	2.5
Copper	6.8
Potassium	1.9
Calcium	3.0
Aluminium	11.8

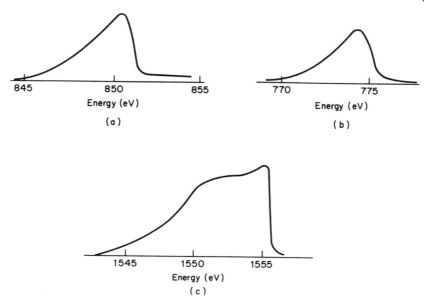

Figure 4.19. Micro-densitometer traces of photographs of the X-ray spectra of solids: (a) nickel; (b) cobalt; (c) aluminium

transition, provided all transitions are equally probable. The densitometer recordings therefore indicate the distribution of states throughout the band. This result is of great importance, since it provides evidence that classical statistics should not be used when the behaviour of electrons in energy bands is being considered, and before proceeding further, it would be worth while to devote a brief section to statistics.

4.8 CLASSICAL AND QUANTUM STATISTICS

Whenever a large assembly of particles is to be dealt with, a statistical approach, including what are termed distribution functions, must be used. If a system contains only a few particles, it is possible to describe the system item by item. For example, it could be said that a given ball on a billiard table was at a given place moving at a given velocity, whilst another was at a different position moving at some other velocity. For a large assembly, such as a volume of gas or a solid containing as many as 10^{28} particles per cubic metre, such detailed knowledge becomes impossible to obtain. Instead, one normally asks such questions as how many particles are contained, say, in a given volume and how many lie within a certain velocity or energy range.

The application of such statistical methods will have been met by most readers in dealing with the simple kinetic theory of gases. This theory deals with a randomly moving set of molecules which are capable of making perfectly elastic collisions with one another. In 1859 James Clerk Maxwell showed that a plot of the number of molecules $n(v)$ having a velocity in the range v to $v + \Delta v$ against velocity would yield a curve of the type shown in Figure 4.20. The equation of this curve (the Maxwell–Boltzmann distribution) has the form

$$n(v)dv = Av^2T^{-3/2}\exp(-Bv^2/T)dv$$

where A and B are constants and T is the absolute temperature. Such a distribution indicates that the majority of molecules have velocities close to a certain mean velocity $\bar{v}$. If the temperature is raised, the maximum shifts to higher energies and the curve flattens out. If the temperature is reduced to zero, the mean velocity also becomes zero, indicating that all the molecules have zero energy at this temperature. The most important assumption of classical statistics, however, is that any number of particles may have identical energies. It has already been shown that an energy band is made up of a large number of discrete levels, each of which can only be occupied by one electron at at time, so it is unlikely that the energy distribution in a band will follow Maxwell–Boltzmann statistics. Enrico Fermi and Paul Dirac in 1926 were the first to obtain the correct distribution of a large number of electrons in a system amongst the available energy levels. They showed that the probability that a particular level of energy E would be occupied at a given temperature was given by

$$F(E) = \frac{1}{\exp[(E - E_F)/kT] + 1} \tag{4.22}$$

where $F(E)$ is called the Fermi function and E_F is the Fermi energy for the particular system, which we met in Section 4.6. Figure 4.21 shows a plot of

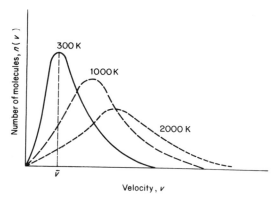

Figure 4.20. Diagram showing how the Maxwell–Boltzmann distribution changes with temperature

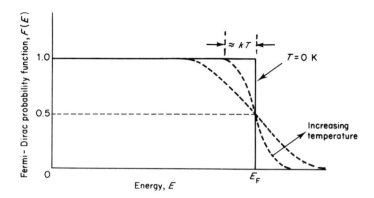

Figure 4.21. The Fermi–Dirac probability function at $T=0$ K and at $T>0$ K

this function for $T=0$ K and $T>0$ K, from which we see that the function is unity for energies less than E_F and zero for energies larger than E_F at $T=0$ K. This means that at absolute zero, all levels below E_F will be completely filled and all those above will be completely empty. Therefore E_F is the energy corresponding to the Fermi level and at absolute zero is the energy of the highest filled level in the band. It is only at $T=0$ K that E_F has an easily identifiable significance, since at higher temperatures the distribution rounds off, as seen from the $T>0$ K curve, and some of the levels below E_F become depopulated and some above become populated. The extent of this rounding off is very slight ($\approx kT$), except at very high temperatures, but it should be noted that, for any finite temperature, $F(E) = \frac{1}{2}$ for $E=E_F$; this means that the Fermi level can also be defined as that level at which the probability of finding an electron is one-half.

The Fermi function does not in itself give the number of electrons that have a certain energy, but only the probability that a given energy state will be occupied by a single electron. To determine the actual number of electrons in the system with a given energy, the number of available energy states within a given range must be known. If the number of states is then multiplied by the probability of occupation, the number of electrons in the system will be obtained. If $n(E)\mathrm{d}E$ is the number of electrons in an energy interval $\mathrm{d}E$ and $Z(E)$ the number of states,

$$n(E)\mathrm{d}E = Z(E)F(E)\mathrm{d}E$$

The function $Z(E)$ is called the state density function, and it can easily be demonstrated that it shows a parabolic variation with energy of the form

$$Z(E) = AE^{1/2}$$

The number of electrons $n(E)$ in a given range $\mathrm{d}E$ is therefore given by

$$n(E)\mathrm{d}E = \frac{AE^{1/2}\,\mathrm{d}E}{\exp\left[(E-E_F)/kT\right] + 1} \tag{4.23}$$

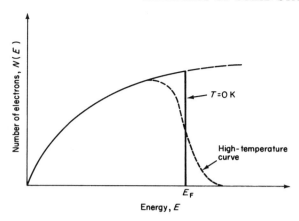

Figure 4.22. The theoretical distribution of the electrons in an energy band

This distribution, known as the Fermi–Dirac distribution, is shown in Figure 4.22 for $T = 0\,\mathrm{K}$ and for $T > 0\,\mathrm{K}$. At absolute zero all states up to the Fermi level are filled, but at higher temperatures the rounding off of the Fermi function leads to a corresponding spreading out towards higher energies of the distribution curve.

Neglecting certain irregularities, the general shape of the densitometer tracings obtained in the soft X-ray experiments is similar to the theoretical curve of Figure 4.22, thereby providing convincing experimental evidence for the validity of Fermi–Dirac statistics.

4.9 CLASSIFICATION OF SOLIDS ACCORDING TO BAND THEORY

This chapter concludes with a consideration of the general classification of solids according to their particular energy band arrangement. Such energy level diagrams for solids can be derived by long and involved calculations, which have been largely substantiated by means of one or more of the modern techniques mentioned in Section 4.7. However, for the purposes of discussion of the electronic properties of solids, it is sufficient simply to quote the final results.

The results of such calculations and experiments on various materials are summarized schematically in Figure 4.23, from which it can be seen that two broad divisions exist; at absolute zero there are solids in which the valence band is full and separated from the conduction band by a definite energy gap and those in which the valence band is only partially filled or overlaps a partially filled conduction band. To the former group belong insulators and semiconductors,

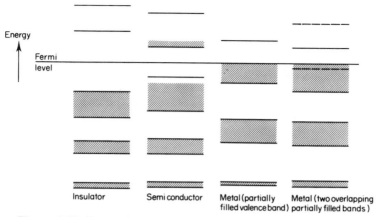

Energy

Fermi
level

Insulator   Semiconductor   Metal (partially   Metal (two overlapping
                                      filled valence band)   partially filled bands)

Figure 4.23. Energy band scheme for various classes of material

and to the latter group metals. The feature distinguishing insulators and
semiconductors is the width of the energy gap. In insulators E_g is of the order
of 4.0 eV and in semiconductors it is typically 1.0 eV, so in semiconductors there
is a possibility of excitation of electrons into the conduction band, even at room
temperature. It might be thought that a divalent metal would have a completely
filled valence band and therefore behave like an insulator. However, owing to
band overlap as shown in Figure 4.23, metals such as calcium and magnesium
are conductors.

4.10 SUMMARY

In this chapter we have considered the concept of energy bands somewhat in
isolation. However, as we shall see in the subsequent two chapters, the power of
the theory lies in its application to specific problems in solid state physics, such as
in the quantum theory of conduction and the operation of semiconductor devices.

PROBLEMS

4.1 Show that the energy levels and wave functions for a particle moving in
the xy plane in a two-dimensional box of side lengths a and b are

$$E = \frac{\hbar^2}{2m}\left(\frac{n_1^2}{a^2} + \frac{n_2^2}{b^2}\right)$$

and

$$\Psi = C\sin\left(\frac{n_1\pi x}{a}\right)\sin\left(\frac{n_2\pi y}{b}\right)$$

Discuss the degeneracy when $a = b$.

4.2 Calculate the amount of energy emitted when an electron in a three-dimensional box makes a transition from the second excited state to the ground state. Assume that the box is cubic of side length 0.5 nm.

4.3 If the energy in the above problem appeared as radiation, calculate the corresponding wavelength.

4.4 A particle with zero potential energy is confined within an infinitely deep one-dimensional well of width d. Show that the energy is quantized and that the wave functions have the form:

$$\psi = \sqrt{\frac{2}{d}} \sin\left(\frac{n\pi x}{d}\right)$$

Sketch the solutions for $n = 1$ and $n = 2$.

4.5 The density of states function for electrons in a metal is given by

$$Z(E)dE = 13.6 \times 10^{27} \, E^{1/2} dE$$

Calculate the Fermi level at a temperature a few degrees above absolute zero for sodium which has 2.5×10^{28} electrons per cubic metre.

4.6 Using the results of the previous problem calculate the velocity of electrons at the Fermi level in sodium.

4.7 Sketch the first three Brillouin zones for a two-dimensional lattice of constant a. Calculate and comment on the area of each zone.

5 Quantum Theory of Conduction

The fundamental assumption of the Drude kinetic theory was that each particle in the electron gas tended to move against an applied electric field whilst still obeying classical statistics. This implies that each electron is capable of taking on any value within the energy range of the distribution. As was shown in Chapter 4, however, the electrons in a solid cannot take on every possible energy value. A continuous range of energies is not available and a set of allowed energy bands occurs, each band being separated from the next one by a forbidden energy gap. Each allowed energy band comprises a set of very closely spaced energy levels, each of which, at any instant, can be occupied by a single electron or be unoccupied. The energy bands are filled up with electrons, the lower energy value bands being occupied by electrons close to the nuclei of the atoms making up the solid. The electrons with energies corresponding to these bands play little part in determining the electrical properties of a solid and are often ignored. Eventually when all of the electrons within the solid have been accommodated in energy bands, there will be upper energy value bands which may be completely full, partially filled or completely empty, and which correspond to the valence and conduction bands mentioned in Chapter 4. The behaviour of electrons with energies corresponding to these two bands is crucial in determining the nature of the solid. We shall now consider the behaviour of electrons in completely and partially filled bands.

5.1 THE BEHAVIOUR OF ELECTRONS IN ENERGY BANDS

5.1.1 FILLED BANDS

Consider a solid (for example an insulator) whose valence band is completely filled with electrons. In the absence of an applied electric field, there is no net movement of the electrons as a whole in any specific direction and hence no net current flow. If an electron does move and hence acquires extra energy, and consequently moves into a higher energy level, another electron will replace it immediately. This is simply a manifestation of the Pauli principle, which, it will be recalled, does not allow more than one electron to have a given set of quantum numbers (or energy coordinates); that is not more than one electron can occupy a single energy level within a band. Electrons in a filled band, therefore, may constantly interchange their energy levels without any net increase in energy.

Suppose that an electric field is now applied to the solid. All of the electrons in the valence band, and indeed those in lower energy bands, will experience a force, which is proportional to the field but which acts in the opposite direction. If a valence band electron were to move as a consequence of the applied field it would take energy from the field and change its energy state, that is move to a higher energy level in the valence band. This is not possible, however, since at any instant all the energy levels are, on average, occupied. There are therefore no higher energy levels available for the electron to move into without violating the Pauli principle. The electrons cannot increase their energy and therefore in general cannot move in the direction opposite to that of the field; that is they cannot acquire a drift velocity and there is thus no net current flow. We see, then, that if the valence band is completely filled, the electrons cannot give rise to a current flow despite there being a very large number of them. Solids whose valence bands are completely full and which have the next highest energy band, the conduction band, completely empty are therefore insulators. Solids comprised of atoms with even numbers of electrons, for example carbon, are often insulators.

5.1.2   PARTIALLY FILLED BANDS

The situation with partially filled bands is considerably different and depends on whether the band contains only a few electrons, whether it is approximately half filled or whether there are only a few remaining empty levels near the top of the band.

Half-filled bands

When the band is half filled, it can contribute greatly to the conduction because there is a large number of electrons and a large number of available higher energy states into which they can go when a field is applied. This situation is characteristic of metals. For example, in sodium, which was discussed in Section 4.5.1, it was seen that the highest occupied band corresponds to the 3s electron subshell which can accommodate up to two electrons. Thus a piece of sodium metal comprising N atoms will have $2N$ available energy levels in the uppermost occupied energy band. As sodium is monovalent, however, there are only N electrons to be accommodated in the band, which is thus half full.

In this case, if a given electron moves to a new, higher energy level in the band as the result of the application of a field, Pauli's principle is no longer violated and the electron can accept energy from the field and move through the crystal lattice. Indeed, many electrons can move in this way and an electric current results. It should be noted that although the electrons in this band are the valence electrons of the constituent atoms, it is often referred to as the conduction band, because current flow in the metal results from the motion

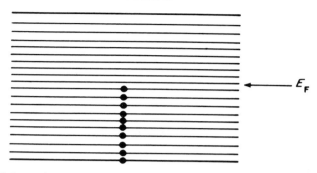

Figure 5.1. Schematic representation of the electrons in the conduction band of a monovalent metal. In the absence of applied electric fields, and at low to moderate temperatures, the levels within the band are occupied up to (approximately) the Fermi energy E_F

of electrons in this band. Although these electrons are often referred to as free electrons, they are not free in the sense of the free electrons in the classical electron gas. The free electrons in this 'quantum' electron gas are subjected to the distribution of available energy levels within the band. In other words, the electrons do not obey classical statistics (which would allow them to acquire any energy within a continuous range of energies), but quantum statistics (which restricts them to certain specific energy values).

It is perhaps worth emphasizing this latter point by remembering that an energy band comprises a very large number of energy levels, which, though very closely spaced, are discrete. We also remember that only one electron can occupy a given level at any given instant. The distribution of electrons in the absence of an external field is then as illustrated in Figure 5.1, the highest occupied level in the band being approximately at the Fermi energy (or level), E_F, which was introduced in Section 4.6 (at absolute zero the upper occupied level would coincide with the Fermi level). If a field is now applied it is evident that only the electrons nearest the top of this distribution will be able to respond easily. Referring to Figure 4.21, it might be expected that only those electrons within about an energy kT of E_F will respond and behave like classical free electrons.

Incidentally, if the temperature of the metal is raised, only the electrons within about kT of the Fermi level move to higher energy levels. This point explains the absence of a significant contribution to the specific heat capacity of metals. It will be remembered that in Section 1.2 we estimated the specific heat capacity of solids by associating $\frac{1}{2}kT$ with each degree of freedom possessed by the atoms in relation to both the kinetic and the potential energies of their vibrations. For a monovalent metal with as many free electrons as there are atoms, a contribution of $3 \times \frac{1}{2}kT \times N$ due to the kinetic energy of the free electrons might have been expected. The absence of this 'electron specific heat' is explained quantum mechanically by noting that only a fraction, of the order of kT/E_F, of the electrons are actually able to change their energy if the temperature changes. This is in contrast to the change in velocity (that is energy) when the temperature of a collection of atoms, which obey Maxwell–Boltzmann statistics, is raised, where from Figure 4.20 it can be seen that most of the atoms will experience an increase in energy.

Example 5.1

Calculate the actual electron specific heat in copper.

Solution. The Fermi level in copper is about 7.0 eV, while kT at room temperature is 0.025 eV. Hence in a kilomole of copper the number of electrons n that contribute to the specific heat is

$$n = N \times 10^3 \times \frac{kT}{E_F}$$

The energy of these electrons, on the basis of equipartition of energy, is

$$E = N \times 10^3 \times \frac{kT}{E_F} \times \frac{3}{2} kT$$

$$= \frac{3}{2} \frac{kT}{E_F} RT$$

The resulting electron specific heat is thus

$$C_{ve} = \frac{dE}{dT} = 3 \frac{kT}{E_F} R = 1.1 \times 10^{-2} R$$

This contrasts with the value of $\frac{3}{2}R$ predicted on the basis of a classical electron gas.

On the basis of the above discussion, magnesium, which is divalent (it has two 3s subshell electrons), could have been expected to have an upper occupied energy band that is completely full; that is for a piece of magnesium of N atoms, there are $2N$ valence electrons, which will fill the band so that magnesium might be expected to be an insulator. In fact magnesium and several other solids with divalent atoms display typically metallic properties and are usually referred to as metals. We shall return to the case of the divalent metals below, and in Section 5.1.3.

Nearly empty and nearly full bands

Energy bands which have either a few occupied levels near the bottom or a few unoccupied levels near the top of the band are of considerable interest. One of the reasons for this is the theoretical predictions (supported by experimental measurement) that the masses of the electrons in energy bands differ from the normal electronic rest mass, vary with position in the energy band and indeed may even in the uppermost energy levels be negative. This was discussed in

Section 4.5.3, where it was shown that the electron effective mass is given by

$$m_e^* = \frac{\hbar^2}{d^2 E / dk^2} \qquad (4.21)$$

The basic reason for this behaviour is not too difficult to understand. We are, of course, discussing the behaviour of electrons *inside* solids. Thus an electron that moves in response to an external electric field will be subjected to a variety of interactions with other electrons and the ion cores as it moves through the solid. It should come as no surprise, therefore, if its response is different from that of an electron in a vacuum subjected to the same field. To account for these internal interactions we simply use Newton's second law to write the acceleration of the electron as

$$a = \frac{F}{m_e^*} = \frac{e\,\mathcal{E}}{m_e^*} \qquad (5.1)$$

where m_e^* is the *effective* mass of the electron.

It is useful to invoke another idea that enables us to refine the energy band model. It will be recalled that Hall effect measurements (see Table 3.4) for certain metals including cadmium indicated that positive charge carriers seemed to be responsible for the electrical conduction; the same is true for some semiconductors. These results are in fact obtained for solids with nearly full and nearly empty energy bands. These and related conceptual difficulties can be removed by introducing the idea of a 'positive hole'.

5.1.3 POSITIVE HOLES

Suppose one is standing on the stage of a theatre in which the stalls are completely filled and the circle is completely empty. A person is seen to go up from the stalls to the circle to occupy one of the many vacant seats there, leaving an empty seat behind. This is then occupied by a former neighbour who, having moved, leaves behind in turn another empty seat. A third person now moves to occupy this, so that the whole process repeats itself. This problem could be approached from two angles: either the individual movements of all the people present could be investigated or only the motion of the vacant seat could be considered (that is either a many-body problem, which is exceedingly complex, or a single-body problem, which is relatively simple, could be dealt with). Clearly the latter is the more desirable, and one need only think of the empty seat moving downwards as the person moves upwards, and moving to the left as all the individuals move to the right, as illustrated in Figure 5.2.

Again, if the model of an energy band that consisted of a flask filled with water (Section 4.6) is recalled, the concept of a hole can be appreciated even further. Suppose a bubble is introduced at the bottom of the flask and is seen to rise to the surface. No one would really tackle this problem by considering

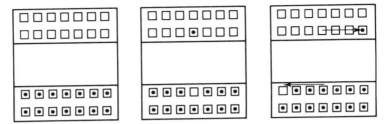

Figure 5.2. Theatre analogy of electron–hole pair

the behaviour of the individual water droplets as they move around in order to fill the volume vacated by the bubble; the obvious procedure would be to concentrate on the movement of the bubble.

In simple terms, then, the hole is the 'absence' of an electron, or a vacant state; that is the people and water droplets are replaced by electrons and the empty seats and bubbles by 'positive holes'. The situation existing in nearly filled or almost empty energy bands can now be appreciated.

If a band has N available states, of which M are occupied, then

$$N - M = p$$

is the number of unoccupied states or holes that can carry current. The symbol p (for positive) is always used to designate holes. The holes can then be treated in their own right, remembering that they carry positive charges since the removal of a negative charge leaves behind a positive one.

Holes can be considered as particles that can undergo collisions, drift and behave in general just as electrons do. It will be recalled that in the last section it was mentioned that electrons near the top of the band have a negative effective mass. It is now evident that the resultant behaviour of the electrons in a band from which a few electrons have been removed (from the top of the band) is most readily explained by considering the behaviour of the corresponding positively charged holes with positive effective mass.

On the other hand, the behaviour of electrons in bands in which there are only a few occupied levels is often quite well described by the free electron model, as there are very many empty states into which these electrons can move as a result of an applied external field or rise in temperature. We would, of course, have to use the appropriate effective mass in calculating the acceleration and drift velocity.

The concept of how holes can arise in a solid is most easily illustrated by considering the energy band scheme of an intrinsic semiconductor. The term intrinsic semiconductor is often used to denote a pure semiconductor, in contrast to extrinsic semiconductors (see Section 5.2.1), which have impurities deliberately added to them. Figure 5.3 shows the energy bands of a typical intrinsic semiconductor, for example silicon, with an energy gap, E_g, of about 1 eV (that

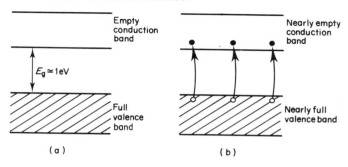

Figure 5.3. Creation of electron–hole pairs in an intrinsic semiconductor: (a) at $T=0\,K$; (b) at $T>0\,K$

is 1.6×10^{-19} J). Thus, except at absolute zero when the conduction band is completely empty, there is a finite probability that some electrons will acquire sufficient thermal energy to jump from the valence band to the conduction band. This leads to a nearly full (valence) band and a nearly empty (conduction) band. From Figure 5.3(b) it is clear that when an electron makes the transition, an empty level or 'hole' is left in the valence band. Electron excitation can also occur as a result of an electron in the valence band absorbing a photon of energy $h\nu > E_g$. If an electric field were now applied to such a solid, electrons in the conduction band and holes in the valence band would drift in opposite directions and both would thus contribute to the conductivity (in view of their opposite charges), as illustrated in Figure 5.4.

The total conductivity is therefore the sum of that due to the electrons moving within the conduction band and that due to the holes moving within the energy restrictions of the valence band. Thus

$$\sigma_{total} = \sigma_{hole} + \sigma_{electron}$$

or

$$\sigma_{total} = ne\mu_e + pe\mu_h \tag{5.2}$$

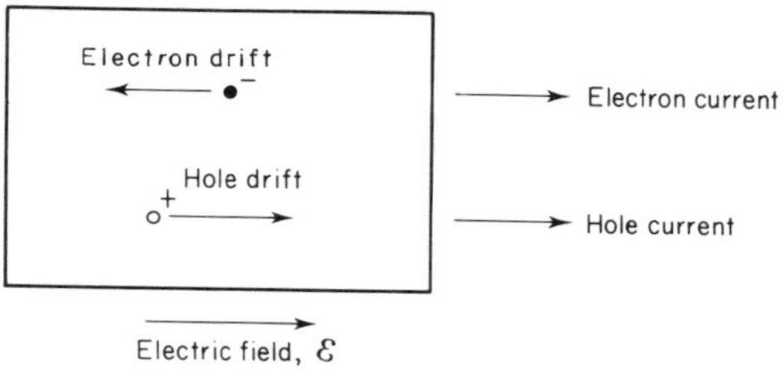

Electron drift

Hole drift

Electron current

Hole current

Electric field, $\mathcal{E}$

Figure 5.4. Electron and hole drift in a semiconductor when an electric field $\mathcal{E}$ is applied

where n and p are the concentrations (that is numbers per unit volume) of electrons and holes respectively and μ_e and μ_h are their mobilities. We need not worry about the opposite signs of the electron and hole charges as the electron and hole conductivities both act in the same 'sense', that is contribute to current flow in the conventional forward direction.

For the intrinsic semiconductor shown in Figure 5.3, the numbers of electrons and holes are equal since there is a one-to-one correspondence between electrons and holes in the two bands and we often write $n = p = n_i$ so that

$$\sigma_i = n_i e (\mu_e + \mu_h) \tag{5.3}$$

where the i denotes intrinsic.

Example 5.2

Calculate the conductivity of intrinsic silicon, for which, at room temperature, $n_i = 1.6 \times 10^{16} \, \text{m}^{-3}$, $\mu_e = 0.14 \, \text{m}^2 \, \text{V}^{-1} \, \text{s}^{-1}$ and $\mu_h = 0.05 \, \text{m}^2 \, \text{V}^{-1} \, \text{s}^{-1}$.

Solution. We have

$$\sigma_i = 1.6 \times 10^{16} \times 1.6 \times 10^{-19} \times (0.14 + 0.05)$$
$$= 4.86 \times 10^{-4} \, \Omega^{-1} \, \text{m}^{-1}$$

This value is rather less than that of metal conductors such as copper ($\sigma_{Cu} \approx 10^8 \, \Omega^{-1} \, \text{m}^{-1}$) but significantly greater than that for good insulators (for example glass, $\sigma_{glass} \approx 10^{-12} \, \Omega^{-1} \, \text{m}^{-1}$), thereby justifying the term semiconductor.

While in intrinsic semiconductors $n = p$, in general this is not the case; indeed, in extrinsic semiconductors we usually find that either $n \gg p$, or $p \gg n$, (Section 5.2.1). The dominant carrier type can usually be determined from a measurement of the Hall coefficient. We saw in Section 3.3.3 that this parameter can be written as $R_H = 1/ne$, so that for electrons R_H is negative. If there are both electrons and holes present then equation (5.3) is rewritten as

$$R_H = \frac{n\mu_e^2 - p\mu_h^2}{e(p\mu_h + n\mu_e)^2} \tag{5.4}$$

Clearly if $n \gg p$, this equation approximates to equation (3.10). It is clear from equation (5.4) that the sign of the Hall coefficient will give the sign of the carriers which are present in the greatest concentrations. This will not necessarily be the case, however, if n and p are nearly equal, as the relative magnitudes of μ_e and μ_h will then also have to be taken into account.

Before leaving this topic let us return briefly to the dilemma concerning the divalent solids such as cadmium and zinc. We recall that these solids (a) have positive Hall coefficients and (b) might be expected to have a filled uppermost occupied band and thus be insulators, but in fact display properties that are essentially metallic in nature. It is difficult to explain these observations completely using one-dimensional theory (which is essentially what we have used

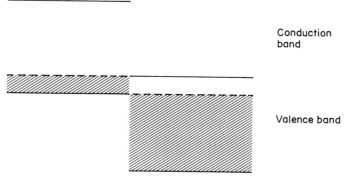

Conduction band

Valence band

Figure 5.5. Schematic representation of the effects of overlapping bands in, for example, divalent metals. Electrons from the top of the lower band occupy states in the upper band, thereby creating both electron and hole carriers

so far); we really need to move at least to two dimensions and to use the concept of Brillouin zones (introduced in Section 4.5.1). This we shall not do, but rather attempt to give a qualitative explanation as follows. Metallic conduction is characterized by the ready availability of a large number of unoccupied higher energy levels, as, for example, in a half-filled band. Such a situation could exist in cadmium if the uppermost energy bands were similar to those in insulators or semiconductors, except for the fact that the conduction band and valence band overlap as shown in Figure 5.5. Some electrons would then move from the valence band into the conduction band (as the energy at the bottom of the conduction band is lower than that at the top of the valence band), leaving holes behind. The concentration of electrons in the conduction band must be the same as that of the holes in the valence band, so the sign of the Hall coefficient, from equation (5.4), depends also on the relative magnitude of the mobilities μ_e and μ_h.

The behaviour of electrons in energy bands can be further illustrated by an extension of our water analogy. Let us imagine a number of sealed glass tubes, one of which is empty, the next has a small amount of water in it, while the next is almost full of water and the last is completely full. If we lie these tubes horizontally they represent respectively an empty energy band, a partially filled band, a nearly full band and finally a completely full band. Let us now tilt the tubes slightly by raising the right-hand ends, which is equivalent to applying an electric field from the left to the right of the tube.

Clearly, in the empty tube there is no movement of water, that is there is no resultant current flow from empty bands. The water in the partially filled tube runs easily down to the left, that is there is current flow to the right. In the case of the nearly full tube, the most obvious effect is that of the small air space at the top of the tube moving to the right. Thus the motion of all of the water to the left results in the empty space, that is the holes, moving to the right in the direction of the applied field. Finally, in the completely full tube there is no movement of the water, that is there is no current flow in filled bands.

5.1.4 MEAN FREE PATHS

We have seen that the quantum theory of conduction applied via the energy band model has enabled us to explain a number of aspects of the behaviour of solids. Another problem to be resolved is that of the exceedingly long mean free paths, λ, that electrons appear to have in solids. How does the quantum theory of conduction help to explain the observation that the value of λ is typically one hundred times that expected on the basis of the Drude theory? The clue lies in the fact that the band theory is a development of the theory applied to the free atom to derive its allowed electron energy levels. In Section 4.2 the wave equation was used with certain boundary conditions to determine the permitted energy states. Thus, under certain circumstances and in order to explain certain experimental facts, an electron must be described in terms of a wave. This can help to account for the longer path lengths.

According to quantum mechanics, an electron travelling in a crystal can be described as a wave that is modulated by the electrostatic influence of the fixed ion cores that it encounters. As the electron moves near an ion, the electron wave is distorted and shows a set of rapid oscillations, as illustrated in Figure 5.6 for a hypothetical linear array of atoms. These imply that the electron is accelerated into a state of higher kinetic energy as it passes the ion. This, in turn, implies that the electron spends comparatively little time near the ion cores and as a consequence is not greatly influenced by them. Indeed, in many solids the conduction electrons can be regarded as free particles as Drude suggested, though with an effective mass m_e^* different from the mass m of electrons in vacuum (Section 5.1.2), which accounts for the slight interaction that does occur.

The Schrödinger wave equation can be applied to this problem by considering an electron moving through the periodic potential distribution of a row of equally spaced atoms such as described in Section 4.5.2. The solution of the problem indicates that if the potential were *perfectly* periodic the electron would have an infinitely long mean free path, that is it would never be scattered. In practice

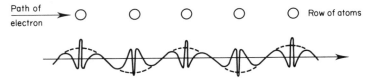

Figure 5.6. Wave function of an electron moving along a row of atoms. The wave function is modulated by the electrostatic influence of the atoms it encounters

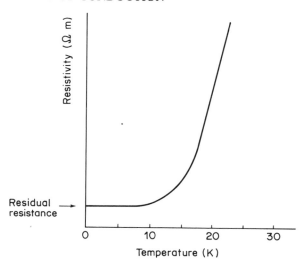

Figure 5.7. Typical variation of the resistivity of a metal with temperature at very low temperatures. The extrapolated value of resistivity at $0\,\mathrm{K}$ is $\approx 5 \times 10^{-11}\,\Omega\,\mathrm{m}$

real solids have imperfections in them which distort the periodicity of the potential distribution resulting from their crystalline nature (see Sections 2.8 and 3.3.4). It is these distortions which act as electron scattering centres and determine the electron mean free path. The imperfections include chemical impurities, physical defects such as vacancies, interstitials, dislocations and grain boundaries, and lattice vibrations (Sections 1.2 and 2.8.4). The latter are caused by the thermal vibrations of the atoms which produce sufficient irregularities to scatter the electrons to some extent. Indeed, the lattice vibrations are responsible for the familiar temperature dependence of the electrical resistance of metals, while the other defects give a residual resistance that is revealed at very low temperatures, as illustrated in Figure 5.7. In semiconductors this temperature-dependent part of the resistance is usually masked by other effects (in particular the exponential increase in electron and hole concentrations with temperature; see Section 5.4).

An alternative approach to explaining the longer-than-expected mean free path is based on the distribution of electrons in the conduction band of a metal, which as we have seen has a maximum at about the Fermi level, E_F. Thus when an electric field is applied, the electrons most likely to respond will be those with energies at the top of the distribution, that is those electrons with energy of the order of E_F. The corresponding velocity of these electrons will then be given by

$$\tfrac{1}{2}mv_\mathrm{F}^2 = E_\mathrm{F} \tag{5.5}$$

where v_F, the Fermi velocity, is much greater than the average thermal velocity calculated from the equipartition of energy; as we can see from Example 5.3, the mean free path, $\lambda = v_F \tau$, is much longer than that given by $\lambda = \bar{v}_T \tau$.

Example 5.3

Evaluate the mean free path in copper for which the Fermi energy is 7.0 eV, given $\tau = 2.4 \times 10^{-14}$ s.

Solution. We have, from equation (5.5),

$$v_F = \left(\frac{2E_F}{m} \right)^{1/2} = \left(\frac{2 \times 7 \times 1.6 \times 10^{-19}}{9.1 \times 10^{-31}} \right)^{1/2} = 1.6 \times 10^6 \text{ m s}^{-1}$$

Hence $\lambda = v_F \tau = 1.6 \times 10^6 \times 2.4 \times 10^{-14} = 38$ nm, which is fairly close to the results quoted in Table 3.5 and much greater than the value given by $\bar{v}_T \tau = 2.88$ nm.

It has been shown that the shortcomings of the classical theory of conduction can be surmounted satisfactorily when quantum ideas are applied to the problem. No claim is made that the above treatment is rigorous, but the reader should bear in mind that all the results can be confirmed using the methods found in more advanced texts on solid state physics. More specific problems, in particular the electronic properties of semiconductors, will now be examined.

5.2 THE ELECTRONIC PROPERTIES OF SEMICONDUCTORS

5.2.1 INTRINSIC SEMICONDUCTORS

In Chapter 4 and Section 5.1.3 the energy band structure for an intrinsic semiconductor has been established and in Section 2.6.2 it was shown that the atomic bonding for typical semiconductors such as silicon and germanium is of the covalent type. Silicon and germanium can be represented schematically by the two-dimensional array of atoms shown in Figure 5.8(a), where each atom is surrounded by four electron pair bonds on account of its tetravalency. It should be remembered, however, that the actual lattice is three dimensional and

(a)   (b)

Figure 5.8. Schematic representation of an intrinsic semiconductor lattice: (a) at absolute zero; (b) at a temperature greater than zero, when some covalent bonds will be broken

that a given silicon atom is surrounded tetrahedrally by four identical atoms with which it shares electrons, thereby completing its outer electron shell. This results in a very stable structure and, owing to the strength of the bonds, there are very few free electrons within it so that, at low temperatures, the electrical conductivity is quite low (Example 5.2). Indeed, at absolute zero silicon would be expected to be an electrical insulator.

Because of the interaction of the atoms in a solid that give rise to energy bands, the valence electrons can be thought of as being shared by the solid as a whole and not localized on specific atoms. At temperatures above absolute zero one can imagine an exchange of electrons between different atomic bond positions. At low temperatures nearly all of the bonds are complete at any instant, but as the temperature rises some of the bonds are incomplete. To put it another way, some electrons have been excited from the valence band into the conduction band, leaving vacant states or holes, which correspond to incomplete bonds, in the valence band, as illustrated in Figures 5.3 and 5.8(b). Due to the constant interchange of valence electrons between atoms the holes also appear to move. Thus the free electrons in the conduction band and holes in the valence band are constantly mobile so that both contribute to the electrical conductivity if a field is applied. The process illustrated in Figure 5.8(b) is usually referred to as the 'breaking' of the covalent bonds. As we have seen, the conductivity of silicon is rather low, and even at moderate temperatures comparatively few bonds are broken to produce free carriers. In fact, as we shall see in Section 5.4, the concentration of free carriers increases exponentially with increasing temperature, resulting in a rapid rise in the conductivity. Another way of increasing the concentration of electrons in the conduction band or of holes in the valence band is, however, rather more generally useful. This is to deliberately add selected impurities to the semiconductors to produce impurity or *extrinsic* semiconductors.

5.2.2   EXTRINSIC SEMICONDUCTORS

By adding minute amounts (typically a few parts per million) of suitable impurities to the melt during the growth of semiconductor crystals it is possible to increase their conductivities. For silicon and germanium, which are tetravalent, that is from group IV of the Periodic Table, two types of impurities are used. These are from groups III and V of the Periodic Table, that is trivalent elements such as boron and indium and pentavalent elements such as arsenic and phosphorus. The reasons for this choice are as follows.

When intrinsic silicon is doped with one of the group V elements, for example phosphorus, each phosphorus atom is found to occupy an atomic site normally occupied by a silicon atom, as shown in Figure 5.9 (phosphorus is a substitutional impurity in silicon). Since the host silicon atoms are tetravalent only four of the five valence electrons of the impurity are used in forming covalent bonds,

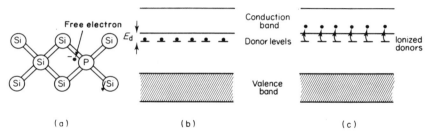

Figure 5.9. An n-type semiconductor: (a) schematic lattice; (b) donor levels at absolute zero; (c) donor levels at room temperature

leaving one electron weakly bound to its parent atom. This electron is easily freed, that is it is easily excited into the conduction band to become a conduction electron.

On the band scheme the energy levels for these extra electrons therefore lie at energy E_d just below the conduction band and are termed *donor levels* since they donate electrons to the conduction band. Correspondingly, the pentavalent impurities are called donors. At absolute zero the donor levels are all occupied (Figure 5.9b), but even at moderately low temperatures most of the electrons are excited into the conduction band (Figure 5.9c) because E_d is so small (≈ 0.03 eV), thereby increasing the free electron concentration and hence the conductivity of the material. A semiconductor of this type is said to be *n-type* since the conduction is predominantly due to negative charges (that is electrons). The electrons are then called *majority carriers* and holes formed by the excitation of electrons from the valence band to the conduction band are *minority carriers*.

It is quite a simple task to estimate E_d. If a donor impurity atom such as phosphorus loses its fifth valence electron it will be left with a net positive charge (the impurity is said to have been ionized). It can be imagined, therefore, that this electron is attached to its parent atom in a situation similar to that found in the hydrogen atom, where a charge of $+e$ binds an electron to the nucleus. The ionization energy of hydrogen atoms is 13.6 eV, but in the case under discussion there are two important differences arising from the fact that the electron moves within a solid. Firstly, the free electron mass m must be replaced by the effective mass m_e^* of the electron. Secondly, the relative permittivity of the semiconductor must be included in the derivation of the equation for the binding energy. This is because the electron orbit is large enough to embrace a significant number of silicon atoms so that it may be considered to be moving in a dielectric medium of relative permittivity ϵ_r. Therefore the excitation energy E_d is given by

$$E_d = 13.6 \times \frac{m_e^*}{m} \times \frac{1}{\epsilon_r^2} \qquad \text{eV} \qquad (5.6)$$

For silicon, for which $\epsilon_r = 12$ and $m_e^* = 0.26\,m$, $E_d = 0.025$ eV.

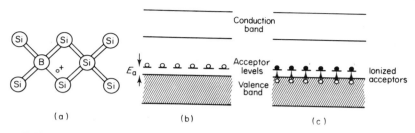

Figure 5.10. A p-type semiconductor: (a) schematic lattice; (b) acceptor levels at absolute zero; (c) acceptor levels at room temperature

Suppose, on the other hand, that silicon is doped with a trivalent impurity such as boron. Again it is found that the impurity atoms occupy sites normally occupied by silicon atoms. In this case it is not difficult to see that there is one electron too few to complete the tetravalent covalent bonds and a vacancy is formed as shown in Figure 5.10. This vacancy is not a hole as it is bound to the impurity atom but, at any temperature above absolute zero, an electron from a neighbouring atom can fill the vacancy, leaving a hole in the valence band. For this reason the trivalent impurities are referred to as acceptors as their associated energy levels accept electrons excited from the valence band. It is often convenient to regard this situation as an acceptor atom with a positive hole orbiting around it analogous to the situation described above. The energy E_a required to 'free' the hole from its parent impurity can be estimated as above, though we would need to use the effective mass of the hole. In reality, of course, E_a is the energy required to excite an electron from the valence band to the energy levels associated with the acceptor impurity atoms. It is found by substituting the hole effective mass into equation (5.6) that $E_a = 0.036$ eV. An impurity atom which loses its associated hole in this way becomes a negatively ionized impurity.

An extrinsic semiconductor with acceptor impurities is referred to as *p-type* since the conduction of current through it is predominantly by positive charges (that is holes). The holes are now the majority carriers and any electrons in the conduction band as a result of excitation from the valence band are minority carriers.

While the elemental semiconductors such as silicon and germanium have been important for a long time, in the last few years several important devices, which use compound semiconductors, have been developed. As the name implies, compound semiconductors comprise two or even three or four elements to form a semiconductor with an appropriate energy gap and other properties. Typical examples are GaAs (gallium arsenide), GaInAs (gallium indium arsenide) and SiN (silicon nitride), which are all used in optoelectronic devices, while GaAs is increasingly being used to fabricate devices for high-speed electronic circuits on account of its relatively high electron mobility (Section 3.2.1).

The question arises as to how GaAs can be doped to make it either n- or p-type. The answer is basically the same as for silicon, that is we add impurities of higher or lower valency than that of the basic host material. In the case of gallium arsenide it is a little more complicated in that the host has two elements, namely gallium (from group III) which is trivalent and arsenic (from group V) which is pentavalent. GaAs is an example of a III–V semiconductor. We can thus make it n-type by, for example, adding tellurium from group VI as a donor impurity or alternatively we can make it p-type by adding zinc from group II as an acceptor impurity. It is interesting to note that the addition of silicon (from group IV) can make gallium arsenide either p- or n-type depending on the conditions obtaining when the material is prepared. The properties of some other important compound semiconductors, including CdS (cadmium sulphide), which is a II–VI semiconductor, are shown in the table in Appendix 1.

The electronic properties of extrinsic semiconductors can be described in terms of their energy band configurations, together with certain interband energy levels introduced by the impurities that are rather sparingly distributed throughout the host material. Another useful parameter in the description of both intrinsic and extrinsic semiconductors, and of the devices fabricated from these materials, is the position of the Fermi level.

5.3   THE FERMI LEVEL IN SEMICONDUCTORS

The Fermi level was defined in Section 4.6 as being the energy level below which all electron states are occupied and above which all states are unoccupied at absolute zero, that is it separates filled states and empty states.

Thus in a metal the Fermi level coincides with the top occupied level in a partially filled band. When the temperature rises, some electrons will move from energy levels below E_F to levels above E_F, thereby conforming with the distribution shown in Figure 4.21. The Fermi level now no longer separates filled and empty states but it is still a useful reference energy. For intrinsic semiconductors (and insulators) at absolute zero we can place the Fermi level somewhere between the valence band (all states filled) and the conduction band (all states empty). It can be shown in fact that the Fermi level is very nearly midway between the bands, that is

$$E_{F_i} \approx \frac{E_g}{2} \tag{5.7}$$

where E_{F_i} is the Fermi level in intrinsic semiconductors and E_g is the energy gap between the valence and conduction bands. When the temperature rises some electrons will be excited from the valence band to the conduction band so there is no clear energy level separating filled from empty states; indeed,

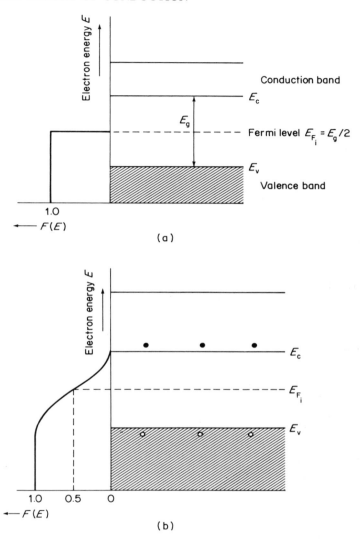

Figure 5.11. (a) Position of the Fermi level in an intrinsic semiconductor $(T = 0\,\text{K})$. (b) The Fermi level E_{Fi} in an intrinsic semiconductor in relation to the distribution function, at a temperature $T > 0\,\text{K}$

we now have both filled and empty states in the two bands. Nevertheless, the Fermi level, positioned at approximately the mid point in the energy gap, still provides a useful reference energy. It is useful to superimpose the Fermi–Dirac probability distribution on the energy band diagram for the cases of $T = 0\,\text{K}$ and $T > 0\,\text{K}$. Figure 5.11(a) shows that at $T = 0$, $F(E) = 1$ for all allowed energy values less than E_{F_i}, while $F(E) = 0$ for all allowed energies greater that E_{F_i}. Figure 5.11(b) shows that for $T > 0$ K there is a small but finite probability of some states in the conduction band being occupied, together with a small but finite probability of some states in the valence band being unoccupied. In intrinsic semiconductors there is always a one-to-one correspondence between electrons and holes so E_{F_i} remains at $E_g/2$ (in fact more rigorous theory shows that E_{F_i} does vary slightly with temperature, and in a given semiconductor depends on the effective masses of the electrons and holes).

Turning now to the case of extrinsic semiconductors let us consider first n-type material with donor levels just below the conduction band as described above. At absolute zero these are the highest occupied levels, so the Fermi level lies between the donor energy levels and the bottom of the conduction band, which of course is empty. In fact E_{F_n}, the Fermi level in n-type material, lies at an energy approximately $E_d/2$ below the conduction band.

As the temperature rises two electron excitation processes occur; that is electron–hole pairs are created by electron excitation from the valence band to the conduction band and also electrons are excited into the conduction band from the donor levels. At low temperatures the latter process predominates as very much less energy is required to cause such a transiton, say 0.03 eV compared to 1 eV for the interband transition. As the temperature continues to rise, however, there is a greater probability that electrons can be excited directly across the energy gap. Since there are many more electrons in the valence band than in the donor levels, such interband transitions increasingly swamp the transitions from the donor levels. Eventually at moderate temperatures ($\approx$ room temperature) virtually all of the donor level electrons have been excited into the conduction band. The only excitation process then remaining is the interband one so that eventually the situation practically reverts to that of intrinsic material at temperatures of about 300°C.

Figure 5.12 shows the energy bands for n-type material with the Fermi–Dirac distribution superimposed, for $T > 0$ K, while Figure 5.13(a) shows the variation in the position of the Fermi level E_{F_n} as the temperature rises.

A similar situation holds for p-type material, although in this case we can imagine that holes are excited into the valence band by two competing processes; in reality, of course, electrons are being excited either into the acceptor levels or conduction band from the valence band. Figure 5.13(b) shows the variation of E_{F_p} with temperature.

When the electronic properties of semiconductors are being discussed the temperature and impurity concentrations must always be carefully stipulated. For many purposes, however, the donor or acceptor levels can be neglected and the behaviour of the material described in terms of the position of the Fermi level. It should be noted that the position of the Fermi level in a given material can be calculated precisely if the impurity concentrations, energy gap and temperature are known. Such calculations confirm the broad details of the above discussion.

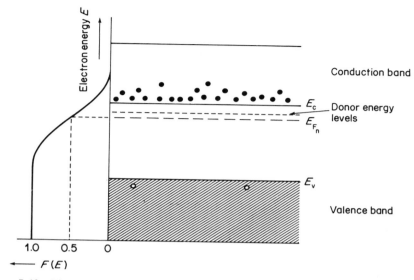

Figure 5.12. The Fermi level E_{F_n} in an n-type semiconductor in relation to the distribution function, at a temperature $T > 0\,\mathrm{K}$

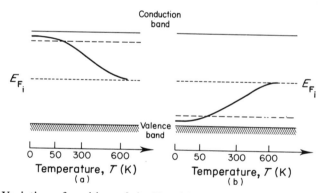

Figure 5.13. Variation of position of the Fermi level with temperature in: (a) n-type semiconductors; (b) p-type semiconductors

5.4 CONDUCTIVITY OF SEMICONDUCTORS

It is clear from equation (5.2) that the concentrations of electrons in the conduction band and of holes in the valence band will influence the value of the electrical conductivity of semiconductors. Using Fermi–Dirac statistics it can be shown that the concentration of electrons in the conduction band is given by

$$n = \text{constant} \times T^{3/2} \times \exp\left[-(E_c - E_F)/kT\right] \qquad (5.8)$$

where E_F is the appropriate Fermi level and the constant term includes the electron effective mass. Equation (5.8) is normally written as

$$n = N_c \exp\left[-(E_c - E_F)/kT\right] \qquad (5.8a)$$

where N_c is a quasi constant as it incorporates the term $T^{3/2}$ and it is therefore only a constant at constant temperature.

For intrinsic semiconductors, $E_F = E_{F_i} \approx E_g/2 \ (\approx (E_c - E_v)/2)$ and therefore

$$n = n_i = \text{constant} \times T^{3/2} \times \exp\left(\frac{-E_g}{2kT}\right) \qquad (5.9)$$

The intrinsic conductivity can thus be written as

$$\sigma_i = n_i e(\mu_e + \mu_h) = \text{constant} \times T^{3/2} \times e(\mu_e + \mu_h) \qquad (5.10)$$

The temperature variation of the conductivity depends both on the temperature variation of the excitation of electrons into the conduction band (equation 5.9) and on the variation of the mobilities, μ_e and μ_h, with temperature. The mobilities at all but the lowest temperatures are largely determined by lattice scattering of the charge carriers and hence, as might be expected, decrease with increasing temperature (the displacements of the atoms from their equilibrium positions increase as the temperature rises). In simple terms the mobilities are expected to vary with temperature as

$$\mu = \text{constant} \times T^{-3/2} \qquad (5.11)$$

though in practice the power of T is often nearer -2 to -2.5.

Substituting equations (5.9) and (5.11) into equation (5.10) gives the following expression for the conductivity of an intrinsic semiconductor:

$$\sigma_i = \sigma_{oi} \exp(-E_g/2kT) \qquad (5.12)$$

where σ_{oi} is a constant.

For extrinsic semiconductors we must take into account the excitation of carriers from the impurity levels on the carrier concentrations. We again use equation (5.8) and substitute the appropriate expression for the Fermi level. For example, for n-type material at low temperature we find that

$$n = \text{constant} \times \exp(-E_d/2kT) \qquad (5.13)$$

while for higher temperatures, where all of the donors are ionized, we have

$$n = N_d \qquad (5.14)$$

(This assumes that the temperature is not so high that the behaviour has reverted to being essentially intrinsic, that is $n_i \not> N_d$.)

In general, the conductivity of an extrinsic semiconductor is made up of two terms, that is

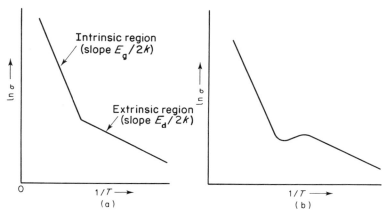

Figure 5.14. Variation of conductivity with temperature for an extrinsic semiconductor for: (a) an ideal case; (b) a more realistic case

$$\sigma = \sigma_{oi}\exp(-E_g/2kT) + \sigma_{oe}\exp(-E_d/2kT) \qquad (5.15)$$

where σ_{oe} is another constant. In practice $E_g \gg E_d$ so that the two terms dominate in different temperature regions. At low temperatures the extrinsic term, $\sigma_{oe}\exp(-E_d/2kT)$, is the most important, while the intrinsic term (equation 5.12) dominates at high temperatures.

The form of equation (5.15) provides a method of measuring both E_g and E_d, for as suggested above one or other term can be neglected depending on the temperature. Thus, if the electrical conductivity is measured as a function of temperature and if $\ln\sigma$ is plotted against $1/T$ a graph of the form shown in Figure 5.14(a) is obtained. The graph consists of two straight line regions, the slopes of which yield E_g and E_d for the high- and low-temperature regions respectively.

Example 5.4

Given the following data for the variation of the conductivity of intrinsic silicon with temperature evaluate the energy gap of silicon:

Temperature (°C)	−23	2	17	42	67
Conductivity ($\Omega^{-1}\,m^{-1}$)	26.7	295	2050	10 800	45 032

Solution. From the data provided the values of $1/T$ and $\ln\sigma$ can be computed. The slope of the graph $\ln\sigma$ versus $1/T$ is then

$$\text{Slope} = 6532$$

which from equation (5.12) equates to $E_g/2k$. Hence the energy gap of silicon is

$$E_g = 1.8 \times 10^{-19}\,J = 1.125\ eV$$

In practice a graph of the form shown in Figure 5.14(b) is obtained where there is a region in which the conductivity decreases as the temperature rises (despite equations 5.9 and 5.13). This occurs in the intermediate temperature range (say 0–60°C for a germanium crystal of moderate doping) where the impurities are all ionized, but the electrons being excited right across the forbidden gap are few in number, so that the carrier concentrations remain almost constant at the value $n = N_d$ as the temperature rises. Therefore the variation of conductivity with temperature is determined by the temperature variation of the mobility which, as we saw above, was of the form $T^{-3/2}$.

Similar arguments apply for p-type semiconductors and E_a can be obtained from appropriate graphs. It is noteworthy that, for semiconductors with reasonably large doping levels, at room temperature the carrier concentrations in n-type and p-type respectively are $n = N_d$ and $p = N_a$ and hence that to a very good approximation we have

$$\sigma_n = N_d e \mu_e \qquad \text{as } n(=N_d) >> p \qquad (5.16a)$$

and

$$\sigma_p = N_a e \mu_h \qquad \text{as } p(=N_a) >> n \qquad (5.16b)$$

In fact it is found that for a semiconductor in equilibrium the product of n and p is given by n_i^2, that is

$$n \times p = n_i^2 \qquad (5.17)$$

Thus if we know n (or p) from the doping concentrations then we can calculate p (or n).

Example 5.5

Given that a piece of n-type silicon contains 1×10^{22} m^{-3} phosphorus impurity atoms, calculate the carrier concentrations at room temperature.

Solution. Phosphorus in silicon is a donor impurity; hence

$$n = N_d = 1 \times 10^{22} \text{ m}^{-3}$$

The intrinsic electron concentration in silicon at room temperature is 1.6×10^{16} m^{-3}; hence, using equation (5.17),

$$p = \frac{n_i^2}{n} = \frac{2.56 \times 10^{32}}{1 \times 10^{22}} = 2.56 \times 10^{10} \text{ m}^{-3}$$

(*Note:* this example demonstrates the validity of equation 5.16a.)

We have now shown that the electrical conductivity of a semiconductor can be increased by heating it or by adding appropriate impurities, as both of these actions lead to an increase in the carrier concentrations. A third method is available for increasing carrier concentrations and hence conductivity, namely the irradiation of the semiconductor with light of sufficiently high frequency. This statement conveniently leads us to a discussion of the optical properties of semiconductors.

5.5 THE OPTICAL PROPERTIES OF SEMICONDUCTORS

In discussing the photoelectric effect in Chapter 1 it was seen that a photon can give up its energy to a single electron near the surface of a solid, thereby enabling it to escape from the surface with a greater or lesser amount of kinetic energy. The energy of the emitted electron for a given photon energy depends in large part on the energy already possessed by the electron in the solid. In particular an electron from near the Fermi level would leave the surface with the maximum energy

$$\tfrac{1}{2}mv^2_{max} = h\nu - \phi \qquad (1.9)$$

where ϕ, the work function, is the energy required to free the electron, or put another way the energy difference between the Fermi level and vacuum level.

The incident photons can also cause emission from lower energy levels such as E_1, as shown in Figure 5.15. In this case the kinetic energy of the emitted electron will be

$$\tfrac{1}{2}mv^2_1 = h\nu - [\phi + (E_F - E_1)]$$

Thus equation (1.9) only applies to electrons being ejected from near the Fermi level $E_1 = E_F$; it is only these electrons that leave with the maximum velocity.

In much the same way a photon may transfer its energy to an electron within the bulk of a solid on which it falls and consequently is absorbed by the solid. Indeed, if this optical absorption for a semiconductor is measured in the visible

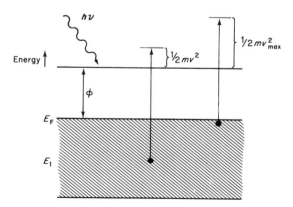

Figure 5.15. Photoelectric emission from a metal exposed to photons of energy $h\nu \geqslant \phi$

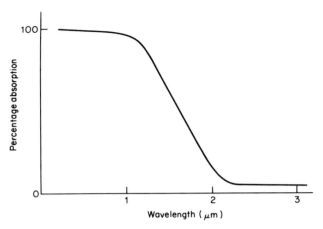

Figure 5.16. The fundamental optical absorption edge

and near infra-red the typical variation of absorption with wavelength is then
as shown in Figure 5.16.

As can be seen, at long wavelengths the absorption is low, that is the material
is highly transmitting until a critical wavelength is reached. At this point there
is an abrupt rise in absorption called the *fundamental absorption edge*. It occurs
at a wavelength typically between 0.7 and 2.0 μm. To appreciate this
phenomenon careful consideration must be given to exactly what is happening
during the recording of an absorption spectrum.

Light from a monochromator is incident on a thin slice of semiconducting
material; the light transmitted by the latter is received and detected by a photocell
whose output is compared with an equivalent signal developed by the incident
beam and then fed to a chart recorder. The wavelength (and hence the energy
of the incident photons) is slowly varied and the fluctuations in absorption
simultaneously recorded. We can understand the mechanism that is taking place
in the sample by comparing it to the analogous situation of the absorption spectra
of gases, in which absorption lines appear when the energy of the incident
radiation happens to coincide with the energy differences between two atomic
energy levels. If we consider what transitions are available in a solid, we are
led to the inevitable conclusion that the transitions between the valence band
and the conduction band must play an important role. The minimum radiant
energy required to bring about such an electron transition can be calculated
from the expression

$$h\nu_g = \frac{hc}{\lambda_g} = E_g \tag{5.18}$$

where E_g is the energy gap, ν_g the associated frequency of the transition and

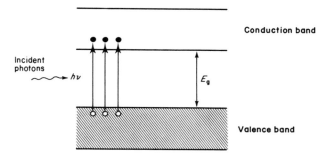

Figure 5.17. Photoexcitation of electrons from the valence band into the conduction band due to the absorption of photons of energy $h\nu \geqslant E_g$

λ_g the corresponding wavelength. Substituting a typical energy gap value of 1.12 eV (for silicon) gives a wavelength of 1.1 μm which agrees with the observed fundamental absorption edge.

If the photon energy is less that $h\nu_g$ then it will not be absorbed whereas if it is greater it probably will be absorbed. While the change from transmission to absorption in some semiconductors, such as gallium arsenide, is almost as sharp as this statement suggests, this is not the case in others including silicon and germanium. The difference arises from the detailed nature of the energy bands of the two groups of semiconductors. Gallium arsenide is an example of a semiconductor having a direct energy band gap, while silicon has an indirect band gap (see Section 5.5.1 below). In the case of semiconductors of the latter type lattice phonons become involved in the optical excitation of electrons from the valence band to the conduction band. In both cases, however, the absorption edge can be measured quite accurately at low temperatures. This provides a convenient way of measuring the energy gap in semiconductors like gallium arsenide, but in the case of materials like silicon some care must be exercised in interpreting the data.

The absorption mechanism is thus clear and is illustrated in Figure 5.17. As soon as the energy of the incident photons equals that of the band gap, transitions can occur between the top of the valence band and the bottom of the conduction band. This amount of energy is thus extracted from the beam, giving rise to a sudden increase in absorption. Clearly, all incident energies greater than this critical value can also cause transitions since there is a very large number of free levels in the conduction band into which electrons may be excited.

One of the obvious consequences of optical absorption in a semiconductor is the creation of additional electron–hole pairs, thereby increasing the conductivity of the material. If an electric field is applied to a sample and the current through it monitored then when the sample is illuminated with light of appropriate frequency a sharp rise in the current is observed. This

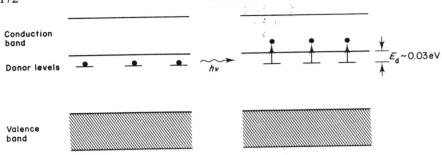

Figure 5.18. Creation of excess carriers in n-type material by photoexcitation: (a) before illumination; (b) during illumination by photons of energy $h\nu \geqslant E_d$

photoconductive effect is of great importance in providing a variety of light-detecting devices such as the CdS and PbS photoconductive cells.

The above description refers in particular to intrinsic semiconductors in which equal numbers of electrons and holes are created. If extrinsic semiconductors are illuminated excess majority carriers of one kind or another are created (depending on the type of material). This is illustrated in Figure 5.18 for n-type material where very long wavelength radiation for which $hc/\lambda_d = E_d (\approx 0.03 \text{ eV})$ will excite electrons from the donor levels into the conduction band. The far infrared absorption spectrum thus shows a small absorption peak centred on E_d. Semiconductors, for example germanium, can be doped with a variety of impurities to produce photoconductive devices sensitive to different wavelength ranges in the infrared part of the spectrum. It is necessary, however, to operate the devices at low temperatures ($\approx 70 \text{ K}$) to ensure that the thermal energy kT available is much less than the impurity ionization energy.

5.5.1 THE HAYNES–SHOCKLEY EXPERIMENT

The creation of electron–hole pairs in extrinsic semiconductors using light was used by J. R. Haynes and W. Shockley (1951) in their celebrated experiment investigating the dynamics of excess *minority* carriers (it will be seen later that the minority carriers often play the most important role in device operation). The basic principles of the experiment, which is shown schematically in Figure 5.19a, are as follows. A pulse of minority carriers, for example holes in an n-type bar of semiconductor, is created by a flash of light incident at some point A near one end of the bar. If the bar has an electric field applied to it, then the pulse of holes will drift along the bar and eventually reach the contact at point C. The presence of these holes will increase the current flowing through the resistor R, and hence the voltage across it. If the oscilloscope, connected across R, is triggered at the instant the flash of light is emitted, then the time t_d taken for the pulse of holes to drift from A to C can be measured from the position of the voltage pulse on the oscilloscope trace (Figure 5.19b).

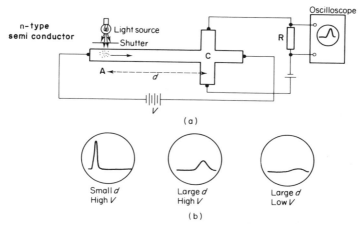

Figure 5.19. (a) The Haynes–Shockley experiment to investigate the dynamics of minority carriers. (b) The oscilloscope trace for various combinations of d and V

This enables us to calculate the drift velocity of the holes, which is simply

$$\bar{v}_D = \frac{d}{t_d}$$

where d is the separation of A and C. If the electric field is $\varepsilon = V/L$, where L is the length of bar, then the mobility of the holes is

$$\mu_h = \frac{\bar{v}_D}{\varepsilon} = \frac{dL}{t_d V} \tag{5.19}$$

It should be realized that the flash of light will create an excess of both minority and majority carriers. It may safely be assumed, however, that the relative increase in the majority carrier concentration $\Delta n/n$ will be very much less than the relative increase in the minority carrier concentration $\Delta p/p$. This is because, although $\Delta n = \Delta p$, n is very much greater than p. Referring back to Example 5.5 it is easy to see that if, for example, $\Delta n = \Delta p = 10^{22}\,\text{m}^{-3}$, then $\Delta p/p \approx 4 \times 10^{11}\,\Delta n/n$. Furthermore, we can ensure that we are only measuring the effects of the excess holes (minority carriers) by making the contact at C a rectifying contact and operating it in reverse bias. This, as we shall see in Section 6.3, only allows minority carriers to cross the contact.

The Haynes–Shockley experiment also provides information on two other very important processes that occur in semiconductors, namely *diffusion* and *recombination*, both of which have important roles in the operation of semiconductor devices (Chapter 6). If the trace on the oscilloscope is observed closely it will be seen that the pulse is much broader in time than the flash of light that generated the carriers. Furthermore, if the pulse is monitored as a function of the drift time t_d, which can be varied either by changing the voltage V or the distance d, it will be seen that the pulse becomes broader as the drift time increases.

The broadening of the pulse is due to the effects of diffusion, which is a movement of charge carriers due to the presence of a concentration gradient. It is important to note that diffusion can and does occur in the absence of any external fields, and is simply the consequence of the random motion of carriers. Let us consider an imaginary plane separating two regions A and B in a semiconductor where the instantaneous electron concentrations are n_a and n_b respectively, with $n_a > n_b$. Then, on average, at any instant the flow from A to B will be proportional to the concentration in region A, that is n_a, while the flow from B to A will be proportional to n_b. There will therefore be a net flow from A to B, that is down the concentration gradient, which is proportional to $n_a - n_b$. Clearly, therefore, the diffusion will tend to smear out the concentration gradient, unless it is maintained in some way. (An everyday example of diffusion is the conduction of heat along a solid from a hot to a cold region. This is due to a diffusion of electrons and/or phonons.)

The diffusion of charge carriers gives rise to a diffusion current, as distinct from drift currents that arise from the presence of a voltage gradient. It seems reasonable to assume that the rate of carrier diffusion is proportional to the concentration gradient. Thus we can write for the electron diffusion current density, J_e,

$$J_e = eD_e \frac{dn}{dx} \tag{5.20a}$$

where D_e is the electron diffusion coefficient. Likewise for holes we have

$$J_h = -eD_h \frac{dp}{dx} \tag{5.20b}$$

where D_h is the hole diffusion coefficient. The different signs in equations (5.20a) and (5.20b) arise from the fact that if the electron and hole concentration gradients are in the same sense, then the electron and hole motions will be in the same sense, and therefore the corresponding currents will be in the *opposite* sense. Contrast this behaviour with the case of electron and hole drift currents!

As we might expect, the diffusion coefficients, which depend on the ability of the carriers to move through the crystal, are related to the carrier mobilities. It can be shown that

$$D_e = \frac{kT}{e} \mu_e \tag{5.21a}$$

and

$$D_h = \frac{kT}{e} \mu_h \qquad (5.21b)$$

These equations are called the Einstein relations.

Returning to the Haynes–Shockley experiment we can now see why the pulse broadens as it drifts along the bar. It is simply a consequence of the holes diffusing outwards from a region of high concentration at the centre of the bunch of excess holes to regions of lower hole concentration—diffusion and drift occur simultaneously.

Let us now consider why the pulse becomes lower in height (more precisely, in view of the diffusion described above, we should consider the area under the pulse on the oscilloscope screen—Figure 5.19b) as the drift time increases. The height of the pulse indicates, of course, the voltage across R, that is the current flowing through it, which is in proportion to the number of excess holes. As the holes drift along the bar some of them will disappear through recombination with the many electrons in the n-type bar. Thus the greater the drift time, the greater will be the amount of recombination that has occurred, and the smaller will be the resulting current through R.

Recombination is a quite complicated process which can occur by several different mechanisms, all of which ultimately result in the annihilation of an electron and a hole. The details of these mechanisms are related to the precise form of the semiconductor energy band scheme, as illustrated by the E versus k curve (Section 4.5). In this respect semiconductors are divided into those in which either direct or indirect recombination predominates as illustrated in Figure 5.20.

In *direct* or *direct bandgap* semiconductors recombination occurs when an electron in the conduction band falls vertically on the E versus k curve into the valence band and recombines with a hole there, usually with the emission of a photon as illustrated in Figure 5.21(a). The energy lost by the electron,

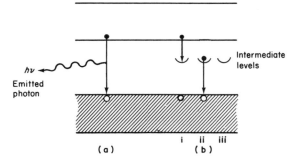

Figure 5.20. (a) Direct and (b) indirect recombination of electrons and holes: (i) the electron is trapped; (ii) the hole is then trapped; (iii) recombination has occurred and the trapping level or recombination centre is unchanged

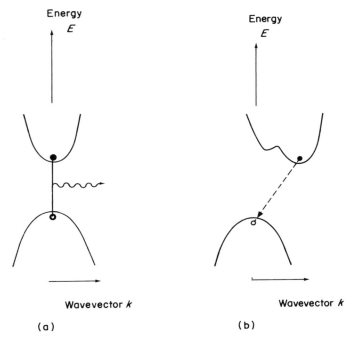

Figure 5.21. Schematic representations of the energy E versus wavevector k curves for: (a) a direct bandgap semiconductor such as GaAs and (b) an indirect bandgap semiconductor such as Si

that is E_g, appears as the photon energy $h\nu_g$, that is $h\nu_g = E_g$. Semiconductors such as GaAs in which this recombination process dominates are used in light emitting devices (Section 6.10.3).

In contrast, in *indirect* or *indirect bandgap* semiconductors such as silicon (Figure 5.21b), the minimum energy points on the E versus k curve for the conduction band and valence band occur at different values of the wavevector k (or momentum p, as $p = \hbar k$). In these cases recombination with the emission of a photon is extremely unlikely as the process somehow has to account for the change in momentum as well as the change in energy of the electron (energy and momentum conservation must be obeyed in quantum mechanical events as well as in classical events). Rather, in these cases, recombination occurs via recombination energy levels, which are localized energy levels occurring in the forbidden energy gap. Any impurity atom or lattice defect can give rise to these levels and act as a recombination centre, if it is capable of receiving a carrier of one type and then subsequently capturing the opposite type of carrier, thereby annihilating the pair. A possible sequence of events is shown schematically in Figure 5.20(b). The first event is the capture of the electron by the recombination centre, with its subsequent transfer to a vacant state, that is a hole in the valence band, after some random time interval. In this process the energy lost by the electron is given up as heat to the crystal lattice; the recombination centre is left in its original state and able to promote another recombination.

It is important to realize that recombination occurs continuously in semiconductors. Indeed, were this not the case, given that thermal generation of carriers occurs continuously at a rate proportional to the temperature, concentrations of carriers would increase steadily; that is for a semiconductor in equilibrium at some particular temperature, the rates of thermal generation and recombination must be equal. It is only when an excess of carriers is created that we can easily study the effects of recombination and, for example, measure the lifetime τ of the carriers. The lifetime is the average time that an electron, having been excited into the conduction band, will remain there before recombining. A similar definition applies to the hole lifetime.

We can measure the minority carrier lifetime using the apparatus shown schematically in Figure 6.35(a). The semiconductor bar is exposed to a steady beam of optical radiation so that excess electrons and holes are created, giving rise to a photocurrent. If the beam is switched off at some instant of time, it is found that the photocurrent shows an exponential decay from which the carrier lifetime can be calculated. We can explain this by assuming that when excess carriers are present the rate of recombination is proportional to the excess carrier concentration, that is

$$-\frac{d}{dt}(\Delta n) \propto \Delta n \qquad (5.22)$$

The minus sign indicates that, as t increases, the excess concentration decreases. We can integrate equation (5.22) to give

$$\Delta n(t) = \Delta n(0)\exp(-t/\tau) \qquad (5.23)$$

where $\Delta n(0)$ and $\Delta n(t)$ are the excess carrier concentrations at $t=0$ and after a time t has elapsed, while τ is the minority carrier lifetime. As we saw previously, we can ensure that we are dealing with minority carriers by using an appropriate rectifying contact.

Returning briefly to carrier diffusion, it is to be expected that as carriers are diffusing they may recombine after, on average, a time τ. The average distance a carrier diffuses before recombining is called the diffusion length L, which is related to the diffusion coefficient and carrier lifetime by the relationships

$$L_e = \sqrt{D_e \tau_e} \qquad (5.24a)$$

and

$$L_h = \sqrt{D_h \tau_h} \qquad (5.24b)$$

Example 5.6

A silicon crystal with a volume of 1 cm³ absorbs 10 mJ of radiation in which the photons are just sufficiently energetic to create electron–hole pairs. Given that the energy gap in silicon is 1.12 eV, calculate the excess carrier concentration.

It is found that the excess carrier concentration falls to 20% of its peak value in 0.1 ms; calculate the carrier lifetime. If the electron diffusion coefficient is 3.5×10^{-3} m s⁻¹, calculate the diffusion length.

Solution. The photons in the radiation must have energy $h\nu = E_g$ hence the number of photons absorbed $= (10 \times 10^{-3})/(1.12 \times e) = 5.57 \times 10^{16}$. Assuming that each photon absorbed generates an electron–hole pair, the excess carrier concentrations are

$$\Delta n = \Delta p = 5.57 \times 10^{16} \times 10^6 = 5.57 \times 10^{22} \text{ m}^{-3}$$

The rate of recombination of the excess carriers is given by equation (5.23) as

$$\Delta n(t) = \Delta n(0) \exp(-t/\tau)$$

Hence $0.2 = \exp(-10^{-4}/\tau)$ and

$$\tau = \frac{10^{-4}}{\ln 5} = 6.2 \times 10^{-5} \text{ s}$$

Finally, the diffusion length is given by equation (5.24a):

$$L_e = \sqrt{D_e \tau_e}$$
$$= (3.5 \times 10^{-3} \times 6.2 \times 10^{-5})^{1/2}$$

or

$$L_e = 4.6 \times 10^{-4} \text{ m}$$

The effect of radiant energy on electrons both near the surface and within a material has now been considered, as has the effect of thermal energy on electrons within the material. We shall conclude this chapter with a brief description of the effect of thermal energy on electrons near the surface, namely thermionic emission.

5.6   THERMIONIC EMISSION

Whilst the vacuum tube is no longer a device of major importance the emission of electrons from hot filaments is still an important phenomenon with many applications. Since electrons do not readily leave metals under normal conditions there must be a potential at the surface of the metal preventing their leaving. The minimum energy that an electron must have in order to escape from the metal is the work function ϕ.

The free electrons inside the metal have an energy ranging from very small values up to E_F. Obviously electrons with energies near E_F which happen to be near the surface have the greatest probability of escaping. Thus if a metal

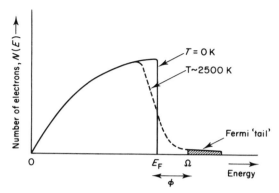

Figure 5.22. The effect of temperature on the Fermi–Dirac distribution

is heated some electrons may acquire the additional amount of energy to enable them to leave the surface and reach the vacuum level, at which point they are considered to be free from the metal. The situation is shown in Figure 5.22. This is equivalent to an atom being ionized when a valence electron is excited to the energy level $n = \infty$, where the electron is considered to be free from the atom. Electrons with energies less than E_F may also be excited to the vacuum level if they acquire sufficient energy. Indeed, electrons at the bottom of the distribution may also be thermally freed from the metal if they acquire an amount of energy Ω, the gross work function; ϕ is often referred to as the net work function and it can be seen that

$$\phi = \Omega - E_F \qquad (5.25)$$

In terms of the Fermi–Dirac distribution, thermal energy causes the discrete edge which exists at $0\,K$ to spread out to form the Fermi 'tail' (Figure 5.22). Those electrons which are excited into the shaded area can be thermally emitted. The origin of the work function can be explained in terms of the energy an electron requires to overcome the net force exerted on it by the positive ion cores tending to pull it back into the metal, as shown in Figure 5.23.

While an electron inside the bulk of the metal experiences very little net force in any given direction, an electron just outside the surface will experience quite a large force pulling it back. This force varies with distance, both because it is Coulombic in nature and because of the increasing number of ions taking part in the interaction as the electron moves further from the surface. The magnitude of the force depends largely on the arrangement of the ions, but also on their charge and the presence or otherwise of surface impurities. Thus the work function varies from metal to metal and even from point to point on the surface of a given metal.

Similar arguments apply also to semiconductors where again the work function is defined as the energy required to excite an electron from the Fermi level to

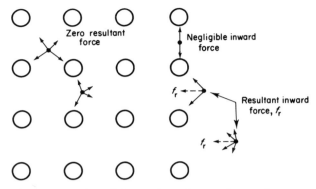

Figure 5.23. Schematic representation of the resultant force acting on electrons at various points within and outside a metal

the vacuum level. A more 'useful' term in the case of semiconductors is the electron affinity χ, which is the energy required to excite an electron from the bottom of the conduction band to the vacuum level—in semiconductors there are not normally electrons at the Fermi level.

As might be expected, the rate of thermal electron emission depends on the temperature. At a given temperature the emission rate is fixed and gives rise to a saturation thermionic current which is expressed by the well-known *Richardson–Dushman equation*

$$J = AT^2 \exp(-\phi/kT) \tag{5.26}$$

where A is constant and ϕ is the work function. It is found experimentally that $\phi \approx 4\,\text{eV}$ for metals (see Table 6.1), but for oxide-coated filaments it is $\approx 2\,\text{eV}$. Thermionic emission is analogous to the situation in which a rocket has to overcome a 'gravitational potential barrier' and reach a critical velocity before escaping from the earth, or in which water molecules have to overcome a surface energy barrier to evaporate from the surface.

5.7 SUMMARY

In this chapter several important electronic properties of solids have been explained in terms of quantum theory and in particular the band theory of solids. Passing reference has been made to the effects on some of these properties of defects in solids, which were discussed in Chapter 2. In the concluding chapter we shall consider the operation of several semiconductor devices, where it will be seen again that the concept of energy bands provides us with an extremely useful model.

PROBLEMS

Use the data provided in the table in Appendix 1 where required.

5.1 Calculate the conductivity of intrinsic germanium and gallium arsenide at room temperature.

5.2 Evaluate the excitation energies E_d and E_a in gallium arsenide.

5.3 The resistivity of intrinsic InAs at room temperature is $3 \times 10^{-4} \, \Omega \, m$. Calculate the intrinsic carrier concentration.

5.4 Measurements of the conductivity of silicon show that in the intrinsic range the conductivity varies with temperature as $\exp(-6493/T)$. Calculate the energy gap of silicon.

5.5 The constant N_c in equation (5.8a) is given by $N_c = 2(2\pi m_e^* kT/h^2)^{3/2}$. Calculate the intrinsic carrier concentration of germanium at 300 K. (The appropriate effective mass to use here is 0.55 m and not the value quoted in Appendix 1).

5.6 The variation of the resistivity of intrinsic germanium with temperature is given by

T(K)	333	385	455	556	714
$\rho(\Omega \, m)$	0.135	0.028	0.0061	0.0013	0.000274

Calculate the energy gap of germanium and estimate the wavelength of the fundamental absorption edge.

5.7 The Hall coefficient of a piece of doped silicon is $+3.8 \times 10^{-4} \, m^3 \, C^{-1}$. Calculate the carrier concentration and, given that the resistivity of the sample is $9.3 \times 10^{-3} \, \Omega \, m$, calculate the mobility of the carriers (state the majority carrier type).

5.8 A piece of gallium arsenide is doped with $1.0 \times 10^{24} \, m^{-3}$ donor atoms. Calculate the carrier concentrations and conductivity. Repeat the problem for germanium.

5.9 A bar of p-type germanium 12 mm long, 1 mm wide and 0.5 mm thick has a resistance of 240 Ω measured parallel to the 12 mm length. Calculate the impurity concentration. Repeat the problem for silicon and in each case estimate the minority carrier concentrations.

5.10 In a Haynes–Shockley experiment on p-type InP the distance between the light source and cross arm was 0.1 m and it was found that the time between the shutter opening and a 'blip' appearing on the oscilloscope screen was 0.25 ms. If the carriers were subjected to a field of 1000 $V \, m^{-1}$ calculate the minority carrier mobility.

5.11 The electron concentration in a rod of silicon of 1 mm² cross-sectional area varies linearly from $10^{22} \, m^{-3}$ to $10^{12} \, m^{-3}$ over a distance of $5 \times 10^{-6} \, m$. Find the electron diffusion current density and current. (Take the temperature to be 300 K.)

5.12 The excess electron concentration in a piece of semiconductor subjected to steady illumination is 5×10^{22} m^{-3}. When the illumination is suddenly switched off the excess electron concentration falls to 2×10^{18} m^{-3} in 25 μs. Calculate the minority carrier lifetime.

5.13 The fundamental absorption edge in a given semiconductor lies at a wavelength of 505 nm and a small absorption peak is also observed at a wavelength of 20 μm. If the Hall coefficient of the semiconductor is negative, draw the energy band diagram of the semiconductor.

6 Semiconductor Devices

In this the final chapter, some of the principles established in the previous chapters will be applied in an endeavour to explain the operation of various solid state devices. A device is defined here as an arrangement that can modify electrical power to produce a useful result. No consideration will be given to the varied electronic circuits in which the devices can be used, but the reader is referred to some of the excellent books available on this topic, a list of which appears on page 243.

Rather, we shall concentrate on the physics of the devices in question and, in particular, use band theory to explain their operation. A convenient starting point for discussing device physics is the formation of a junction between two dissimilar materials.

6.1 METAL–METAL JUNCTIONS

Consideration of what happens when two dissimilar metals are brought into contact provides a particularly useful first example, since the processes that take place can be applied quite generally to junctions between materials of any type. The two metals will be characterized by different band schemes and, in particular, as they have different Fermi levels they will have different work functions. Once again a modified version of the water model (see page 136) can be used, as shown in Figure 6.1. Two tanks A and B are connected by a short length of piping and isolated from one another by a valve (Figure 6.1a). The level of the water (representing the Fermi level) differs in each tank, so that

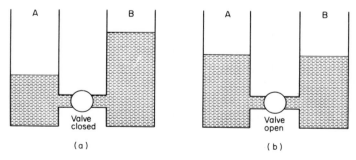

Figure 6.1. Water analogy for Fermi level equalization: (a) with the valve closed; (b) with the valve open

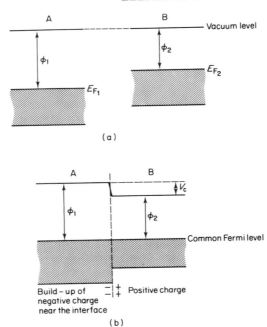

Figure 6.2. Formation of metal–metal junction: energy levels: (a) before junction is made; (b) after junction is made

when the valve is opened water flows from B into A until the levels are equal (Figure 6.1b). In other words, tank A has a deficiency of higher energy drops and tank B an excess, and flow will continue until the uppermost drops in each tank are at the same (potential) energy level.

This model provides a guide as to what will happen initially when two dissimilar metals are brought together. Before the junction is made, the energy band scheme for the two separated metals is as shown in Figure 6.2(a). Since the work functions are different, the Fermi levels are situated at different positions below the vacuum level. When the metals are brought together, therefore, the higher energy electrons in B will flow into A to fill some of the empty lower states until the Fermi levels are equal. Since electrons are removed from B a depletion layer is formed and B acquires a positive charge. Conversely, the additional electrons acquired by A form an accumulation layer and A becomes negative. In other words, the surface of A will be at a different electrical potential from the surface of B, and a potential difference is established across the junction when contact is made. This is called the equilibrium contact potential difference (V_c) and equals the difference in the work functions (expressed in volts); that is,

$$V_c = \frac{\phi_1}{e} - \frac{\phi_2}{e} \qquad (6.1)$$

This charge separation establishes an electric field that hinders the flow of electrons from B to A and encourages a flow from A to B. It is important to note, however, that in equilibrium, although the two metals have a potential difference V_c between them, there is no net current flow. Rather the magnitude of V_c is such as to ensure that there is a dynamic equilibrium between electrons flowing across the junction of the metals in both directions. The equalization of the Fermi levels and of the current flow across the junction between any two materials in equilibrium is generally true, and forms a basis for discussing any junction.

Before discussing a junction it is important to know V_c, since it is this parameter that governs current flow across the junction. The standard method of measuring contact potential differences in metal junctions is the Kelvin method, which has been modified by Zisman (1932) and Riviere (1957). The apparatus is shown schematically in Figure 6.3.

The metals whose contact potential is to be measured are evaporated under vacuum in the form of thin films (A and B) on to two glass plates, which are arranged so as to form a parallel-plate capacitor. Facilities are provided for vibrating one of the plates towards and away from the other, which remains fixed. The plates are connected in an external circuit incorporating an oscilloscope, which detects any current which may pass through the resistor R. When the switch S_2 is closed, metal A is connected to metal B, so current flows until the Fermi levels equalize. This will be seen as a displacement and return to zero of the spot on the oscilloscope screen. A contact potential now exists between the metals, and one plate has become positively and the other plate negatively charged.

The situation is similar to that of a charged parallel-plate capacitor connected to a constant-voltage supply. If the plates are brought closer together, the

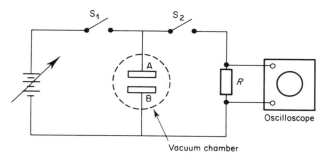

Figure 6.3. Kelvin method for determining contact potential differences of the metals A and B

Table 6.1. Contact potential differences

Metal	V_c (volts)
Ag–Cu	+0.16
Ag–Al	−0.08
Ag–Au	+0.50
Ag–Ni	+0.63
Ag–W	+0.17
Ag–Mo	−0.11

capacitance increases and charge must flow to maintain equilibrium, since the potential between the plates remains equal to V_c. When A is brought towards B, a current flows through R, and if this is repeated rapidly by using some form of vibration, an alternating voltage will be displayed on the oscilloscope. A variable voltage, opposite in sign to V_c, is now applied from a second external circuit by closing switch S_1 and its value adjusted until the alternating waveform on the screen reduces to a straight line. The value of the applied voltage must then equal the contact potential difference.

The reason for performing the experiment under vacuum is to minimize the effects of impurities and oxides on the surface of the film which can considerably alter the work function. Table 6.1 shows some values of contact potential differences for various metal–metal combinations.

In practice, one of the electrodes of the Kelvin apparatus is common to all experiments, so that if the work function of this standard electrode is known, the work functions of all the others may be derived by adding algebraically the known work function to the measured contact potential difference. If a silver electrode is used as standard $\phi = 4.33$ eV, the following values are obtained for the other metals: copper, 4.49 eV; aluminium, 4.25 eV; gold, 4.83 eV; nickel, 4.96 eV; tungsten, 4.5 eV; molybdenum, 4.22 eV. It should be said that although these values are typical there is a wide range of values quoted in the literature.

Although the most familiar example of metal–metal junctions is the thermocouple, it should be stressed that temperature measurement using such junctions does not depend directly on the contact potential at the junction. The phenomenon responsible for producing a potential difference, which varies with temperature, is the *Seebeck effect*.

It was discovered that a temperature gradient within a conductor produces a potential difference between its ends. This arises because energetic charge carriers, electrons or holes, tend to diffuse from the hot end to the cold end of the conductor. Consequently, the ends of the conductor become oppositely charged, as illustrated in Figure 6.4. The electric field produced by the charge separation gives rise to an electric current, which opposes the diffusion current induced by the temperature gradient. An equilibrium is soon established in which

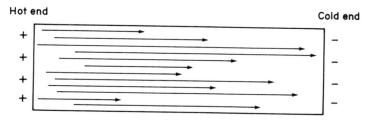

Figure 6.4. Origin of the thermoelectric power in a uniform conductor subjected to a temperature gradient. The arrows indicate schematically the movement of electrons from the hot end of the bar to the cold end, thereby creating an electric field along the bar

these currents exactly balance. The potential difference which develops is called the thermoelectric e.m.f., and is characterized by the thermoelectric power or Seebeck coefficient α_s, which is defined as the potential drop per degree temperature difference:

$$\alpha_s = \frac{dV}{dT} \tag{6.2}$$

For metals in which electrons carry the current, α_s is negative and typically $10\text{--}20 \, \mu V \, °C^{-1}$. For semiconductors, the sign of α_s depends on that of the

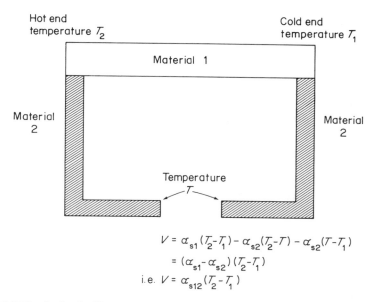

$$V = \alpha_{s1}(T_2 - T_1) - \alpha_{s2}(T_2 - T) - \alpha_{s2}(T - T_1)$$
$$= (\alpha_{s1} - \alpha_{s2})(T_2 - T_1)$$
i.e. $V = \alpha_{s12}(T_2 - T_1)$

Figure 6.5 The Seebeck effect, showing how the thermoelectric voltage V is determined by the difference of the thermoelectric powers of the materials 1 and 2

Table 6.2. Values of α_{s12} (μV K^{-1}) for common thermocouple combinations at room temperature, compared to the values for some pure metal combinations

Copper–constantan	42.7
Chromel–alumel	41.0
Platinum–10% platinum, rhodium	6.4
Iron–constantan	54.9
Silver–gold	0.4
Copper–iron	12.2
Gold–platinum	7.8

majority carriers, and it may be several hundred microvolts per degree Celsius in magnitude. The coefficient α_s is an intrinsic property of the material, but it can never be measured directly since a voltmeter and leads are required to measure the thermoelectric potential. The leads will have the same temperature drop across them as that in the materials under test, so the measured potential difference will be that characteristic of the material reduced (or increased, depending on signs) by that characteristic of the leads. Thus, as indicated in Figure 6.5, it is only possible to measure differences in thermoelectric powers, that is:

$$\alpha_{s12} = \alpha_{s1} - \alpha_{s2} \qquad (6.3)$$

Absolute values of α_s can be obtained indirectly by the use of one material of known α_s, such as that for superconductors, where $\alpha_s = 0$.

Thermocouples comprise pairs of dissimilar metals. Some of the commonly used combinations are listed in Table 6.2 together with the corresponding values of α_{s12}.

Apart from the Seebeck effect there are two other thermoelectric effects, namely the Thomson and Peltier effects. The Thomson effect refers to the heating or cooling that results when an external potential difference drives a current through a conductor in which there is a temperature gradient. Again the Thomson effect is an intrinsic property of the material. On the other hand, the Peltier effect refers to the heating or cooling which arises at a junction between two dissimilar materials when a current is passed through the junction (it must be emphasized that the heating effects mentioned here are additional to any joule heating that might arise). The Peltier coefficient Π is defined as the rate of heat absorption (or emission) per unit current flow, that is

$$\Pi = \frac{\mathrm{d}W}{\mathrm{d}I} \qquad (6.4)$$

It is found that Π and α_s are related by

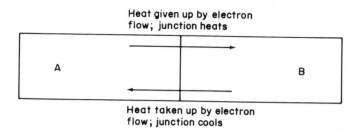

Figure 6.6. The Peltier effect. Electrons in material A have higher energies on average than those in material B. Heating or cooling of the junction then results from electron flow A→B and B→A respectively

$$\alpha_s = \frac{\Pi}{T} \tag{6.5}$$

and as Π may be expressed in terms of fundamental parameters of a metal or semiconductor, we have an alternative way of evaluating α_s.

Whether a junction heats or cools as electrons, for example, flow across the junction depends on the relative energies of the electrons on the two sides of the junction and on the direction of flow. Thus in a double junction ABA between materials A and B, the thermal effect at the junction AB will be opposite to that at junction BA. Let us suppose in Figure 6.6 that the electrons in material A have higher energies than those in material B. Then electrons flowing from A to B will release heat to the lattice, producing a heating effect. Conversely, if the flow is from B to A then the more energetic electrons will tend to leave B, producing a cooling effect. In fact the discussion is rather simpler in terms of applying external potentials to pn semiconductor junctions, which are described later in this chapter.

The Peltier effect is used as the basis for some refrigerators and heat pumps. The best materials for these thermoelectric applications are the semiconductors bismuth telluride (Bi_2Te_3) and antimony telluride (Sb_2Te_3). For example, using these materials a compact refrigerator measuring $10\,\text{mm} \times 10\,\text{mm} \times 5\,\text{mm}$ can pump some $10\,\text{W}$ of heat for a current of $2\,\text{A}$.

6.2 METAL–SEMICONDUCTOR JUNCTIONS

When a metal and a semiconductor are brought into contact the same equilibrium criterion holds as for metal–metal junctions, that is the Fermi levels equalize. This is illustrated in Figure 6.7 for the cases of n-type material forming a junction with metals of both larger and smaller work functions. The work function in a semiconductor is defined in the same way as for a metal, that is, it is the energy separation between the Fermi level and the vacuum level.

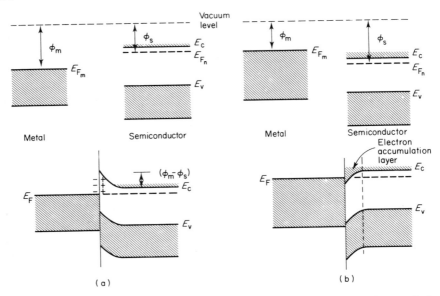

Figure 6.7. Band structure of metal to n-type semiconductor junctions, before and after contact: (a) for $\phi_m > \phi_s$, a rectifying contact is formed; (b) for $\phi_m < \phi_s$ an ohmic contact is formed

In semiconductors, we often define another quantity, namely the electron affinity χ, which is the energy required to take an electron from the bottom of the conduction band to the vacuum level. Thus it has a more clearly defined physical meaning as being the energy required to free an electron from the semiconductor. Remember that even in p-type semiconductors there are some electrons in the conduction band, situated (usually) at the bottom of that band, and so only need energy χ (eV) to release them.

Consider the case of n-type material brought into intimate contact with a metal which has a greater work function ϕ_m than that of the semiconductor ϕ_s (Figure 6.7a). Electrons flow from the semiconductor into the metal until, in equilibrium, the Fermi level is continuous across the junction. This is brought about by electrons leaving donors in the semiconductor near the junction and entering the metal. This process, as we saw in Section 6.1, creates a depletion layer in the semiconductor. In this case there is a positive space charge residing on the ionized donors, which induces a negative surface charge on the metal. The depletion layer (or barrier layer or space charge region) is so called because it has been depleted of its free carriers, and as a consequence it is highly resistive. The electrons that flow into the metal form an accumulation layer next to the junction. Electron transfer continues until the electric field set up by the dipole layer is sufficiently strong to inhibit further electron movement.

This situation is represented by an upward bending of the energy band edges in the semiconductor to form a potential barrier of height $e(\phi_m - \phi_s)eV$, as shown in Figure 6.7(a) (as electrons in the conduction band move towards the negative charge on the metal near the junction their energy must rise, that is the energy bands bend upwards). There will still be some electron flow from the metal to the semiconductor due to thermal agitation. Similarly, a few electrons in the conduction band of the semiconductor will have enough energy to surmount the potential barrier and flow into the metal. In equilibrium these flows will be equal and no net current will flow across the junction.

If the semiconductor is now made positive with respect to the metal by applying an external voltage V (this is referred to as a bias voltage), the energy bands of the semiconductor are lowered by eV. This has the effect of increasing the barrier height and widening the depletion layer as illustrated in Figure 6.8. This condition is referred to as reverse bias. Under these circumstances electron flow from the semiconductor to the metal is almost entirely prohibited by the very high potential barrier of $[eV + e(\phi_m - \phi_s)]$, but the flow from the metal to the semiconductor remains unaltered. Thus the reverse bias current (conventionally from semiconductor to metal) saturates at a low value of V and is essentially independent of the magnitude of V.

If the polarity of the external bias is changed, so forward-biasing the junction, then the barrier height opposing electron flow from the semiconductor to the metal is reduced by eV. There is now a substantial electron flow in this direction,

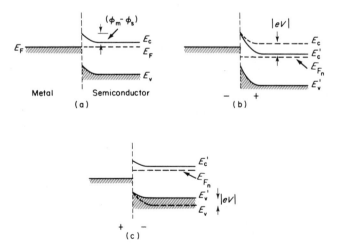

Figure 6.8. Effect of bias on metal to n-type semiconductor junction ($\phi_m > \phi_s$): (a) zero bias; (b) reverse bias; (c) forward bias. When a bias is applied the Fermi levels are no longer aligned. E_{Fn} moves down relative to E_{Fm} with reverse bias and up with forward bias

while again the small flow from the metal to the semiconductor remains constant. There is therefore a large net flow of electrons giving a conventional current from the metal to the semiconductor. It can be shown that this current increases approximately exponentially with the forward bias voltage. Thus the metal–semiconductor junction described passes a large current in one direction but a very small one in the opposite direction, that is it is a rectifier (often referred to as the Schottky rectifier).

If, on the other hand, the metal has a smaller work function than the semiconductor then, as shown in Figure 6.7(b), electrons will flow from the metal to the semiconductor, causing an electron accumulation layer to build up near the junction until the Fermi level becomes continuous in equilibrium. The resulting electric field causes a downward bending of the energy bands in the semiconductor near to the junction. It is clear that in this case there is no potential barrier at the junction and electrons can move freely across it in either direction independent of the sign of the bias voltage. Such a junction is an ohmic junction.

Similar considerations apply to p-type semiconductor–metal junctions, where if $\phi_m < \phi_s$ a rectifying junction is formed while if $\phi_m > \phi_s$ an ohmic junction is formed.

Clearly it is important in making connections to semiconductor devices that ohmic contacts can be formed. The recipe for this is indicated above. Alas, metal to semiconductor junctions rarely behave as predicted and almost invariably rectifying junctions are produced irrespective of the relative sizes of the work functions. This behaviour is due to the presence of states or energy levels at the surface of the semiconductor which can trap charge there, causing band bending even in the absence of any other material. The surface states arise because of the discontinuity in the crystal that is the surface, from the presence of absorbed impurity atoms or oxide layers and from physical defects on the surface. The surface states behave as either electron or hole traps, depending on their origin. Figure 6.9(a) illustrates the band bending which occurs in an n-type semiconductor due to the presence of surface electron traps. The surface states deplete electrons from the bulk material, exposing the donors and creating a depletion layer. This causes a natural potential barrier to occur independently of any metal contact being made. If the density of surface electron traps is high enough electrons may be removed from the valence band as well as the conduction band. The concentration of holes near the surface of the semiconductor may then become larger than the concentration of electrons, in which case the surface layer may become p-type in character. Figure 6.9(b) shows that the bands have bent to such an extent that the Fermi level is nearer to the valence band than the conduction band in a narrow, so-called, inversion layer. The importance of such layers will become apparent when field effect transistors are discussed in Section 6.9.2. For the present discussion, it is obvious that band bending, as described above, or the formation of an accumulation layer, as

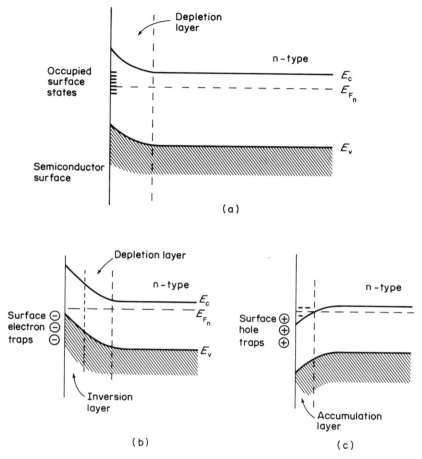

Figure 6.9. (a) Energy band bending due to the presence of surface states on an n-type semiconductor. The surface states, in this example, are occupied by electrons, thereby causing the bands to bend upwards thus forming a depletion layer. (b) The bands in (a) have bent upwards to such an extent that an inversion layer has formed. (c) Formation of an accumulation layer for electrons due to the presence of surface states which trap holes

illustrated in Figure 6.9(c), due to the presence of surface hole traps will significantly affect the behaviour of metal–semiconductor junctions.

Ohmic contacts can, and indeed must, be produced. The techniques used remain somewhat more of an art than a science. An essential step, however, appears to be very heavy doping of the surface region of the semiconductor before applying the metal contact, or indeed during the metallization process. Thus, for example, gold containing a little tin makes a successful contact to n-type silicon. Tin is a donor dopant in silicon, and can thus increase the donor concentration N_d at the surface. Similarly, aluminium makes a good ohmic contact to p-type silicon.

6.3   SEMICONDUCTOR–SEMICONDUCTOR JUNCTION DEVICES

Earlier discussions of conduction processes in semiconductors have been concerned only with uniformly doped single crystals. Such crystals have limited device applications though they are incorporated in photoconductive cells, temperature-sensitive resistors and some bulk effect high-frequency generators. The most useful and interesting devices are those that use semiconductor single crystals in which there is a more or less abrupt change in doping type from acceptors to donors at some plane surface within the crystal. The width of the transition region from one doping type to the other is usually a few tens of nanometres and is usually an important parameter governing the performance of pn junction devices.

6.3.1   THE pn JUNCTION DIODE

The operation of the pn junction diode can be understood by considering the junction to be formed by joining together initially isolated pieces of p-type and n-type semiconductor materials (the practical fabrication of such junctions is discussed in Section 2.10). On making contact, electrons diffuse from the n-region to the p-region because of the concentration gradient that has been established. This depletes the n-material close to the metallurgical junction, which represents the actual p to n transition, of free electrons leaving uncovered (ionized) donors which form a positively charged depletion layer. Similarly, holes diffuse from the p-region to the n-region, creating a negatively charged depletion layer in the p-material next to the junction. These two charged regions together form the depletion or space charge layer, within which the total stationary charge is zero. For this reason the relative widths of the two parts of the depletion layer depend on the relative doping levels, and the depletion layer will extend furthest into the least heavily doped side of the junction, that is $x_n N_d = x_p N_a$, where x_n and x_p are the widths of the depletion layers in n- and p-regions respectively. The carriers which diffuse across the junction become minority carriers and an excess concentration of minority carriers is created. This in turn leads to an excess minority carrier concentration gradient becoming established, so that the minority carriers continue to diffuse away from the junction gradually recombining with majority carriers in the bulk regions as they do so.

The diffusion of carriers across the junction continues until the charge separation and associated electric field in the depletion layer is large enough to prevent (in effect) further diffusion (see Section 6.3.2). At this stage equilibrium

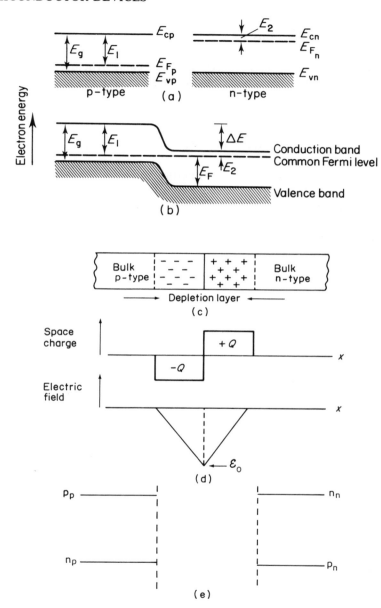

Figure 6.10. The pn junction in equilibrium: (a) p- and n-region energy bands before the junction is formed; (b) energy bands at the pn junction; (c) the junction depletion layer; (d) the space charge and electric field within the depletion layer; (e) carrier concentrations in the bulk regions at the edges of the depletion layer

has been established and there is no net flow of current across the junction. As in the case of the junctions discussed above, equilibrium corresponds to the Fermi levels being aligned across the junction. Since the Fermi level in each material is fixed relative to its respective energy band structure, alignment of the Fermi levels can only be achieved by the energy bands in the p-region rising relative to those in the n-region. This is reasonable as the diffusion of the majority carriers has resulted in the p-region becoming negative with respect to the n-region, and thus an electron in the conduction band of the p-region will have a greater energy than an electron in the conduction band of the n-region.

The energy bands of the two regions before and after contact are shown in Figure 6.10, which also shows the space charge, carrier density, electric field and potential distributions. The equilibrium electrostatic potential V_0, shown in Figure 6.10, between the two regions is called the contact, diffusion or built-in potential. Its magnitude, together with the width of the depletion layer, is largely determined by the doping concentrations. Typically for silicon pn junctions these parameters have values of 0.8 V and 0.1 μm respectively.

Example 6.1

Calculate the equilibrium contact potential difference and depletion layer width for a pn junction in silicon, given that the impurity concentrations are $N_a = 1 \times 10^{23}$ m^{-3} and $N_d = 1 \times 10^{22}$ m^{-3} respectively, while $n_i = 1.6 \times 10^{16}$ m^{-3} at $T = 300$ K and $\epsilon_r = 12$.

Solution. V_0 and the depletion layer width W for an abrupt pn junction are related to the impurity concentrations by

$$V_0 = \frac{kT}{e} \ln \left(\frac{N_a N_d}{n_i^2} \right) \tag{6.6}$$

and

$$W = \left[\frac{2\epsilon_0 \epsilon_r}{e} (V_0 - V_b) \left(\frac{1}{N_a} + \frac{1}{N_d} \right) \right]^{1/2} \tag{6.7}$$

where ϵ_r is the relative permittivity and V_b is the bias voltage applied to the junction (V_b is negative for reverse bias and positive for forward bias — see Section 6.3.3).

From equation (6.6) we have

$$V_0 = \frac{1.38 \times 10^{-23} \times 300}{1.6 \times 10^{-19}} \ln \left(\frac{1 \times 10^{22} \times 1 \times 10^{23}}{2.56 \times 10^{32}} \right)$$

$$= 0.75 \text{ V}$$

From equation (6.7), we have in equilibrium, that is zero bias,

$$V_b = 0$$

and

$$W = \left[\frac{2 \times 8.85 \times 10^{-12} \times 12 \times 0.75}{1.6 \times 10^{-19}} \left(\frac{1}{1 \times 10^{23}} + \frac{1}{1 \times 10^{22}} \right) \right]^{1/2}$$

$$= 0.33 \ \mu m$$

It should be noted that while V_0 is quite real and can be measured experimentally, it cannot be used to perform useful work, as work cannot be obtained from an equilibrium situation. We should also note that, as the depletion layer has lost most of its free carriers, the material there has become very much more resistive than the, so-called, bulk material on either side of it. The contact potential difference, V_0, and any externally applied bias voltage (see Section 6.3.3) will then be essentially 'dropped' across the depletion layer rather than the bulk p- and n-regions. Consequently, the corresponding electric field across the depletion layer is very large while the fields across the bulk regions are very small. It was seen in Example 6.1 that the width of the depletion layer W was of the order of 10^{-6} to 10^{-7} m, so the electric field across it will typically be $10^7 \ \text{V m}^{-1}$, which is an extremely large field.

6.3.2 THE pn JUNCTION WITHOUT AN APPLIED BIAS

As outlined above, no net current crosses an unbiased pn junction even though there is a potential difference between p- and n-regions. This is because, although four current components flowing in the transition region must be considered, the total current crossing the junction under equilibrium conditions is zero. For electrons there is a diffusion current i_e (conventionally from the p- to the n-side) due to the small fraction of the electrons in the conduction band of the n-side, which have sufficient energy to surmount the potential barrier V_0 and cross to the p-side. There is also a minority carrier drift current i_e' (conventionally from the n-side to the p-side) due to the potential gradient from the n- to the p-side. In equilibrium these components are equal and opposite. Similarly there is a hole diffusion current i_h from the p- to the n-side and a hole drift current i_h' from the n- to the p-side; again these components are equal and opposite.

For the purposes of these discussions recombination and generation within the depletion layer will be ignored. This is reasonable in view of the very narrow width of the depletion layer (see Example 6.1) and the short time carriers take to cross it.

Let us firstly consider free electrons in the conduction band of the p-region. These are in continual random motion owing to their thermal energy and move through the crystal within the range of energies imposed by the conduction band. Those electrons that move close to the junction are caught by the electric field there and swept into the n-region. The number of such electrons depends on the concentration of electrons in the conduction band on the p-side, which in turn depends on the temperature of the crystal; that is, as in general the electron concentration in the conduction band is given by

$$n = N_c \exp[-(E_c - E_F)/kT] \qquad (5.8a)$$

the electron concentration in the p-region of the junction will be

$$n_p = N_c \exp[-(E_{cp} - E_{F_p})/kT] \qquad (6.8a)$$

where E_{F_p} is the Fermi level in the p-region. From Figure 6.10 we may write

$$n_p = N_c \exp(-E_1/kT) \qquad (6.9a)$$

as

$$E_1 = E_{cp} - E_{F_p}$$

The electron drift current i_e' will be proportional to this electron concentration, so

$$i_e' = A' \exp(-E_1/kT) \qquad (6.10)$$

where A' is a constant. Similarly, the electron diffusion current will depend on the electron concentration n_n in the conduction band in the n-region and on the probability that these electrons might acquire an energy ΔE and be able to climb the potential barrier between the n- and p-regions. This probability is given by Boltzmann statistics as $\exp(-\Delta E/kT)$. Hence, as

$$n_n = N_c \exp[-(E_c - E_{F_n})/kT] \qquad (6.8b)$$

$$= N_c \exp(-E_2/kT) \qquad (6.9b)$$

we have

$$i_e = A \exp(-E_2/kT)\exp(-\Delta E/kT) \qquad (6.11)$$

where A is another constant and $E_2 = E_{cn} - E_{F_n}$. In equilibrium, however, $i_e' = i_e$; therefore from equations (6.10) and (6.11),

$$A' \exp(-E_1/kT) = A \exp(-E_2/kT)\exp(-\Delta E/kT)$$

$$= A \exp[-(E_2 + \Delta E/kT)]$$

From Figure 6.10 it can be seen that

$$E_1 = E_2 + \Delta E$$

Hence $A' = A$ and therefore

$$i_e' = i_e = A \exp(-E_1/kT) \qquad (6.12)$$

Similarly, for the hole drift and diffusion currents it can be shown that

$$i_h' = i_h = B \exp(-E_1/kT) \qquad (6.13)$$

where B is another constant of the same form as A. The total drift current $i_e' + i_h'$ equals the total diffusion current $i_e + i_h$, and is zero in equilibrium.

6.3.3   THE pn JUNCTION WITH AN EXTERNAL APPLIED BIAS

The balance of drift and diffusion currents described above can be upset if a bias voltage is applied across the junction. As with metal–semiconductor junctions, a semiconductor–semiconductor junction can be reverse- or forward-biased, and consideration will now be given to the effect this has on the current flow across the junction.

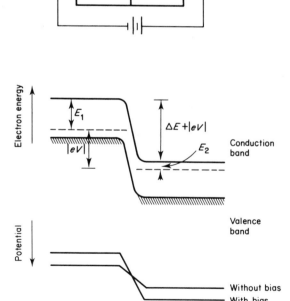

Figure 6.11. The reverse biased pn junction, showing the energy bands and the potential distribution at the junction

For reverse bias a voltage source is connected across the junction such that the p-side is negative and the n-side positive, as shown in Figure 6.11. As described in the last section, the depletion layer is very resistive compared to the bulk regions and consequently the contact potential V_0 together with any external voltage effectively appears only across the depletion layer. The effect of the bias is to raise the energy levels of the p-type side by making them more negative and thereby increasing the electron energy. The potential barrier height therefore increases by eV where V is the applied bias potential.

Even fewer electrons are now capable of moving from the conduction band in the n-side to the p-side, but in the other direction the current will remain very nearly the same as in the unbiased junction. As a result i_e and i_h both become very small. The concentration of electrons in the conduction band of the n- side is $N_c \exp(-E_2/kT)$ (equation 6.9b) as before, but now the probability of them overcoming the potential barrier is reduced to $\exp[-(\Delta E + eV)/kT]$.

Example 6.2

Compare the probabilities of electrons in the conduction band of a semiconductor acquiring energies ΔE and $\Delta E + eV$ above the bottom of the conduction band, for a reverse bias voltage of 2 V.

Solution.   The ratio of the probabilities is given by

$$\frac{\exp\{[-(\Delta E + eV)]/kT\}}{\exp(-E/kT)}$$

or $\exp(-eV/kT)$, which for $V = 2$ V and $T = 300$ K is 2.5×10^{-34}!

The diffusion and drift currents may therefore now be written respectively as

$$i_e = A \ \exp(-E_2/kT)\exp[-(\Delta E + eV)kT]$$

$$= A \ \exp[-(E_2 + \Delta E)/kT]\exp(-eV/kT)$$

$$= A \ \exp(-E_1/kT)\exp(-eV/kT) \tag{6.14}$$

and

$$i'_e = A \ \exp(-E_1/kT) \tag{6.15}$$

Hence from these equations (6.14 and 6.15) it follows that $i_e = i'_e \exp(-eV/kT)$.

The net electron current is then the difference of the diffusion and drift currents, that is

$$I_e = i_e - i'_e = i'_e[\exp(-eV/kT) - 1]$$

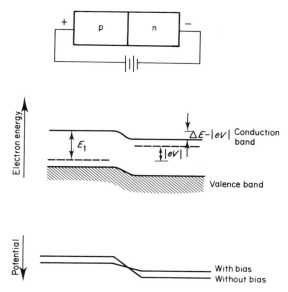

Figure 6.12 The forward biased pn junction, showing the energy bands and the potential distribution at the junction

An analogous expression can be derived for the hole drift and diffusion currents, so that the total current flow can be written as

$$I = I_e + I_h = I_0 [\exp(eV/kT) - 1] \qquad (6.16)$$

where $I_0 = I_e + I_h$ is called the reverse bias saturation current since for values of reverse bias voltage greater than about $-4kT/e$ the exponential term is vanishingly small and the total current saturates at a value of $-I_0$, as shown in Figure 6.13.

When the p-type material is made positive with respect to the n-type, the junction is said to be forward biased. The potential barrier is then reduced by an amount eV, as shown in Figure 6.12. This obviously implies that very many more electrons can flow from the n-side to the p-side than in the unbiased junction, so that the diffusion current is greatly increased. The drift current again is very nearly unchanged when the forward bias is applied, so that now there is a net electron flow from the n-side to the p-side (that is conventional current flows from the p- to the n-side). In a manner completely analogous to that for the reverse bias junction, it can be shown that the current flowing across the junction is again given by equation (6.16), but in this case V is positive. Equation (6.16) is often called the rectifier or Shockley equation. We see that when V is positive, the exponential term dominates and the current rises exponentially. Equation (6.16) is plotted in Figure 6.13 for different values of

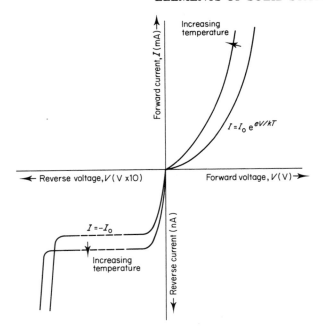

Figure 6.13. Typical characteristics for a silicon pn junction (note the change in scale of the current axis)

temperature, the curve being characteristic of a rectifier. The effect of temperature on the characteristic is also illustrated in Figure 6.14. The difference in the two parts of the characteristics is not surprising since under reverse-bias conditions the device presents an effectively high resistance to the flow of conventional current and under forward bias conditions, a low resistance. As can be seen from Figure 6.13, the main result of a temperature rise is to increase the reverse saturation current as more and more electrons are excited into the p-side conduction band, thereby increasing the drift current. The same is also true for the n-side, and at higher temperatures a larger forward current flows for a given forward voltage. The rapid rise in current at a certain value of reverse voltage will be discussed in the next section; it is often referred to as reverse bias breakdown.

Apart from reverse breakdown the Shockley equation is in good agreement with the experimental data for germanium pn junction diodes at low currents. It is in less good agreement with the behaviour of silicon and gallium arsenide junctions, however, especially at low forward biases. This is essentially because the possibility of carrier recombination in the depletion layer was ignored.

The value of the reverse bias saturation current depends on parameters such as the electron and hole diffusion coefficients and diffusion lengths, and the

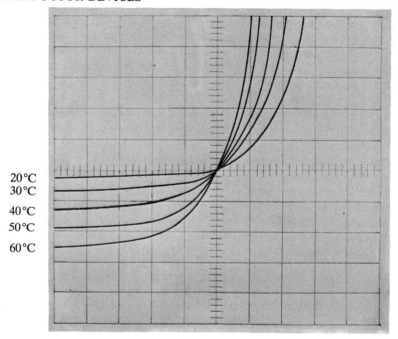

20 °C
30 °C
40 °C
50 °C
60 °C

Figure 6.14. Effect of temperature on pn junction characteristics, at temperatures of 20, 30, 40, 50 and 60 °C: horizontal scale 0.05 V per division; vertical scale 0.01 mA per division

doping concentrations on the two sides of the junction. In germanium I_0 typically varies from a few microamperes down to nanoamperes, while in silicon, typical values range from nanoamperes to picoamperes.

Example 6.3

A silicon pn junction has a reverse bias current at room temperature (22 °C) of 5.0×10^{-11} A. Calculate the current for a forward bias of 0.5 V.

Solution. From equation (6.16) we have

$$I = 5.0 \times 10^{-11} \left[\exp\left(\frac{0.5 \times 1.6 \times 10^{-19}}{1.38 \times 10^{-23} \times 295}\right) - 1 \right]$$

$$= 17.1 \, \text{mA}$$

6.4 REVERSE BIAS BREAKDOWN

It was seen in Section 6.3.3 that at a critical reverse bias voltage, the current through the junction increases sharply, and quite large currents will flow for very little further increase in bias voltage. There is nothing inherently destructive in reverse breakdown, providing the current is limited to a reasonable value by an external resistance. If the current is allowed to become too large then joule heating may damage the junction; the same is of course true if the current becomes excessively large in the forward direction.

Reverse breakdown can occur by two mechanisms, each of which requires a particular critical field across the depletion region. It will be recalled that the depletion region is very narrow so that, even in equilibrium, there is a very high average electric field acting across it. This field may easily be of the order of $10^6 \, \text{V m}^{-1}$, and under reverse bias conditions rise to 10^7 or even $10^8 \, \text{V m}^{-1}$.

The two mechanisms are called the *Zener effect* and the *avalanche effect*. The Zener effect has the higher critical field, but paradoxically breakdown occurs for lower values of reverse bias voltage. This is because the Zener effect predominates in very heavily doped junctions, which, as can be seen from equation (6.7), leads to narrow depletion regions and hence very high electric fields. The avalanche effect occurs in junctions which are more lightly doped, and hence have wider depletion layers.

6.4.1 ZENER BREAKDOWN

When a heavily doped junction is reverse biased, the energy bands become 'crossed' at quite low voltages. By this we mean that the conduction band of the n-side appears opposite, that is at the same energy, to the valence band of the p-side, as shown in Figure 6.15. This has the effect of aligning large numbers of electrons in the valence band of the p-side with empty states in the conduction band of the n-side. There is of course a potential barrier separating these electrons and the empty states, but if the barrier is narrow electrons can cross from the p- to the n-side of the junction by quantum mechanical tunnelling, as illustrated in Figures 6.15(b) and 6.18. (Tunnelling will be described further in Section 6.5.) This may also be imagined as field ionization of the atoms in the depletion layer; that is the electric field across the depletion layer becomes so large ($> 10^8 \, \text{V m}^{-1}$) that it breaks the covalent bonds so that some electrons are torn free from their parent atoms and accelerated to the n-side of the junction. This effect was first discovered in gases by C. Zener in 1934.

6.4.2 AVALANCHE BREAKDOWN

If the depletion layer is wide, as in lightly doped junctions, the electrons and holes being accelerated across this region by the electric field have many opportunities for colliding with the fixed atoms of the lattice. If the energy imparted to the latter is sufficient (that is when the junction field has reached some critical value V_m, which is somewhat greater than $10^6 \, \text{V m}^{-1}$ in silicon), impact ionization may take place and electron–hole pairs are created. In their

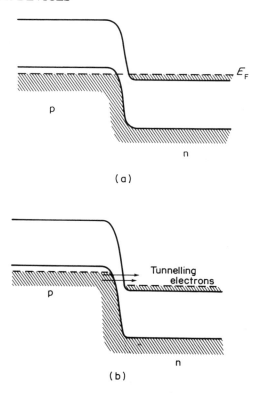

(a)

(b)

Figure 6.15. Tunnel or Zener breakdown: (a) shows a pn junction formed from heavily doped material in equilibrium and (b) shows the junction with a reverse bias voltage applied so that electron tunnelling can occur

turn, these carriers are accelerated in the field and may cause further ionizations and create even more electon–hole pairs and so on. The initial ionization therefore starts an 'avalanche multiplication' process, illustrated in Figure 6.16, which results in a very large current flowing in the reverse direction. If the temperature is reduced, the mean free paths of the charge carriers become longer, so greater energies are imparted between collisions for the same accelerating field, thereby increasing the probability of ionization. The breakdown voltage therefore decreases with decreasing temperature. In contrast Zener breakdown is largely temperature independent. Because of the similarity of the results of these two mechanisms, devices based on them are generally referred to as Zener diodes, but clearly for an avalanche diode this is something of a misnomer. For practical purposes, the actual mechanism giving rise to the particular shape of the characteristic does not matter, and both devices are widely used as voltage reference elements in voltage stabilizer circuits

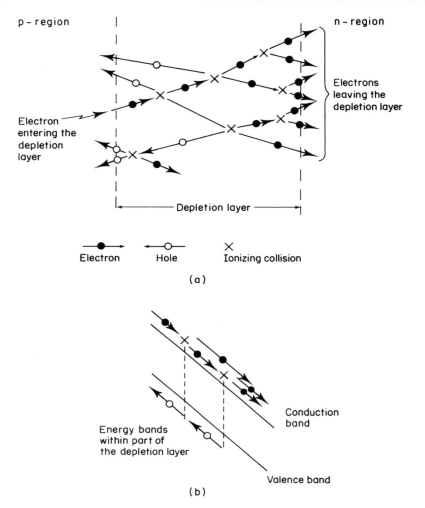

Figure 6.16. Avalanche breakdown: (a) shows the carrier multiplication arising from ionizing collisions in the depletion layer (note that the holes are less effective than electrons in causing ionization and, for clarity, the diagram does not show all of the holes); (b) shows the acceleration of the electrons and holes due to the high electric field in the depletion layer

(a typical Zener diode characteristic is shown in Figure 6.17). By appropriate choice of doping concentrations and method of fabrication, Zener diodes can be manufactured with breakdown voltages ranging from 2–3 V up to about 100 V.

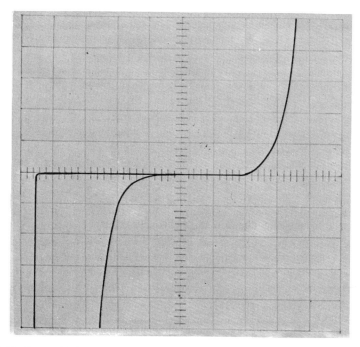

Figure 6.17. Zener diode characteristics: a 9.1 V Zener (in which breakdown is via the avalanche effect) and a 5.1 V Zener (in which breakdown is via the Zener effect). The forward characteristics are indistinguishable. For forward characteristic: horizontal scale 0.2 V per division; vertical scale 0.2 mA per division. For reverse characteristic: horizontal scale 2 V per division; vertical scale 0.2 mA per division

6.5 THE TUNNEL DIODE

The basic characteristics of the tunnel effect can be appreciated by considering the one-dimensional potential barrier shown in Figure 6.18. On the classical picture, an electron with energy $E_0 < V$ would not be able to penetrate the barrier beyond a limit x_1. However, if the wave function in region 1 were to be set up, with suitable boundary conditions (V very large, but finite, for $x_1 < x < x_2$), and the Schrödinger equation solved for region 3, a non-zero solution would be found for region 3. Since the wave function is a measure of whether or not an electron exists in a particular region of space, this can only imply that there is a finite probability that the electron can 'tunnel' through the potential barrier.

This is the situation that exists in a pn junction when the two sides are doped so that the Fermi level actually lies within the conduction band or valence band on the n-side and p-side respectively, and the transition region width is less

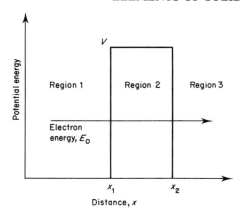

Figure 6.18. The tunnel effect

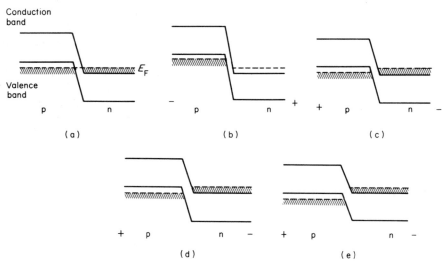

Figure 6.19. Diagram showing the effects of bias on the energy bands of a tunnel diode: (a) zero bias; (b) reverse bias; (c) to (e) progressively increasing forward bias

than $\approx 10^{-8}$ m. An impurity concentration of as high as one part in a thousand may be necessary to achieve this. Such heavily doped materials are said to be degenerate. A device made in this way is known as an Esaki or tunnel diode. Figure 6.19(a) shows that the energy bands of such a heavily doped pn junction are crossed even when no bias is applied. Under these conditions, carriers pass through the transition zone in both directions by tunnelling, but the net current flowing is zero.

When reverse bias is applied (Figure 6.19b), the electrons in the valence band on the p-side 'see' vacant states opposite them in the conduction band on the n-side. They pass through the depletion region by tunnelling and start to occupy the available states. The greater the applied voltage, the more electrons become available for tunnelling as the Fermi level on the p-side becomes higher and higher above that on the n-side. The reverse current then will continue to increase and not reach saturation, as in the normal pn junction.

Various stages under forward bias conditions are shown in Figures 6.19(c) to (e). As the applied forward voltage is increased, the conduction electrons on the n-side find themselves opposite more and more available empty states in the valence band on the p-side and occupy these after tunnelling. Figure 6.19(d) shows the situation when the maximum number of available electrons is opposite the maximum number of available states. In this situation, therefore, the current reaches a maximum and thereafter decreases as fewer states become available for the electrons in the n-side conduction band (whose number is maintained by the power supply to which the junction is connected) to tunnel into. Eventually (Figure 6.19e) there are no more empty states opposite the conduction band electrons, tunnelling ceases and the forward current is carried in the normal way by electrons surmounting the potential barrier between the n-side and the p-side.

If the current is plotted against applied voltage, a characteristic similar to that shown in Figure 6.20 is obtained. The main feature of the tunnel diode characteristic is the negative-resistance section, which finds application in very high frequency (up to 100 GHz) oscillator circuits in the microwave region.

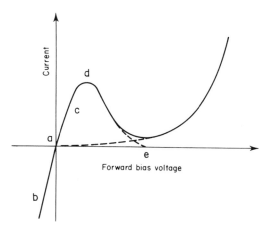

Figure 6.20. Current through a tunnel diode as a function of applied bias; points (a) to (e) correspond to those in Figure 6.19 (a) to (e)

6.6 JUNCTION CAPACITANCE

There are two types of capacitive effect associated with junctions (either pn junctions or metal-semiconductor junctions). The junction or depletion layer capacitance C_j, which predominates in reverse bias, arises from the presence of the two space charge regions comprising the depletion layer. We have seen (equation 6.7) that the width of the depletion layer W, and hence the charge contained within it, varies with the bias voltage, that is, there is a charge varying as a function of voltage which represents a capacitative effect. In truth the junction rather closely resembles a parallel plate capacitor, in which there are two charged conductors, the bulk p- and n- regions, separated by an insulating region, namely the depletion layer. We may therefore write C_j in the form

$$C_j = \frac{\epsilon_0 \epsilon_r A}{W} \qquad (6.17)$$

where A is the cross-sectional area of the junction. This expression is in fact verified by a rather more rigorous approach.

Equation (6.17) shows that the junction capacitance can be controlled by the reverse bias voltage, which alters the value of W. This voltage-dependent capacitance is exploited in the varactor (variable reactor) diode used in tuning circuits and the generation of high frequencies.

The second capacitive effect, which dominates in forward bias, is the charge storage or diffusion capacitance C_d. This effect arises as follows. Let us suppose that a steady forward bias is applied; this will lead to a corresponding level of minority carrier injection. Suppose now that the bias is reduced instantaneously. As a result there will be for a short time a greater number of injected carriers than is appropriate to the new, lower bias; that is, there is a charge storage or capacitive effect. After a short time the additional carriers will have recombined so that the minority carrier concentration will have become appropriate to the new, lower, level of bias voltage. Not surprisingly the value of C_d depends on the steady junction current I, which determines the average carrier concentrations, and the recombination lifetime τ. Thus we have

$$C_d = \frac{eI}{kT} \tau \qquad (6.18)$$

Example 6.4

Calculate the reverse and forward bias capacitances for the pn junction described in Example 6.1, given that the junction has a cross-sectional area of $1 \times 10^{-6}\,\mathrm{m^2}$ and that the carrier lifetime is $\tau = 0.1\,\mu\mathrm{s}$. Assume that the reverse bias voltage is 10 V and that in forward bias a current of 0.1 mA is flowing

Solution.   (a) Reverse bias. With a reverse bias of 10 V, the depletion layer width is

$$W = \left[\frac{2 \times 8.85 \times 10^{-12} \times 12 \times (0.75 \times 10)}{1.6 \times 10^{-19}} \left(\frac{1}{1 \times 10^{23}} + \frac{1}{10^{22}} \right) \right]^{1/2}$$

$$= 1.25 \times 10^{-6} \, \text{m}$$

Therefore,

$$C_j = \frac{8.85 \times 10^{-12} \times 12 \times 1 \times 10^{-6}}{1.25 \times 10^{-6}}$$

$$C_j = 85 \, \text{pF}$$

(b) Forward bias. From equation (6.18) the charge storage capacitance is

$$C_d = \frac{eI\tau}{kT} = \frac{1.6 \times 10^{-19} \times 1 \times 10^{-4} \times 1 \times 10^{-7}}{1.38 \times 10^{-23} \times 300}$$

$$= 386 \, \text{pF}$$

The charge storage capacitance, in particular, can be a serious limitation for forward biased pn junctions in high-frequency circuits. Indeed, whenever a device which includes a pn (or metal–semiconductor) junction is used in alternating current circuits the appropriate junction capacitance combined with the appropriate resistance of the device forms an important time constant. In the case of the pn junction the resistance is a combination of the resistance of the bulk regions, which can usually be ignored, and the junction resistance, which is considered in the next section.

6.7 SLOPE RESISTANCE r_e

Let us consider the resistance seen by the a.c. part of a combined a.c. + d.c. applied potential. If the d.c. potential forward-biases the junction then the changes in current I for small changes in voltage V about a steady bias voltage will depend on the slope of the I–V curve at the particular d.c. bias point; that is a dynamic, differential or slope resistance r_e can be defined such that

$$r_e = \frac{\Delta V}{\Delta I} = \frac{dV}{dI}$$

It should be noted that r_e is not a fixed quantity as the slope of the I–V curve varies as the forward bias changes. Considering the diode equation (equation 6.16)

$$I = I_0 \exp\left[(eV/kT) - 1 \right]$$

we may write

$$\frac{1}{r_e} = \frac{dI}{dV} = \frac{e}{kT} I_0 \exp(eV/kT)$$

or as $\exp(eV/kT) \gg 1$,

$$\frac{1}{r_e} \approx \frac{eI}{kT}$$

or, if I is expressed in milliamperes, then, at room temperature

$$r_e = \frac{kT}{eI} = \frac{25}{I} \ \Omega \tag{6.19}$$

Thus, for a forward current of 5 mA, the slope resistance is only 5 Ω. In reverse bias the situation is rather different as the slope of the I–V curve is very nearly zero, so that r_e tends to infinity; that is

$$\frac{1}{r_e} = \frac{dI}{dV} = \frac{e}{kT} I_0 \exp(eV/kT) \approx 0$$

for $V \gtrsim -4kT/e$. Hence $r_e \rightarrow \infty$. In practice r_e may be about $10^8 \ \Omega$.

6.8   THE BIPOLAR JUNCTION TRANSISTOR

The development of the bipolar junction transistor by J. Bardeen, W. H. Brattain and W. Shockley in 1948 started the modern electronics revolution, which led to a host of electronic systems and enabled computers to be miniaturized. The main functions of bipolar junction transistors are in amplification and switching, processes which will be described in the following sections.

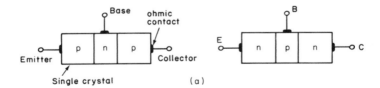

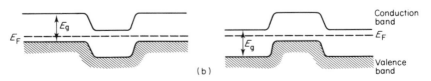

Figure 6.21. Junction transistors: (a) pnp and npn transistors; (b) energy band schemes with zero bias

A junction transistor can be envisaged as being made simply by forming two pn junctions back to back in a single crystal of semiconductor, thereby forming a sandwich structure. Clearly either the p-material or the n-material can constitute the 'filling', so, as Figure 6.21 shows, there are two basic types of transistor, pnp or npn. The npn transistor is usually preferred in silicon since the mobility of minority electrons in the base (see below) of an npn transistor is much higher than that of holes in the base of a pnp device. This leads to higher operating speeds, that is better high-frequency performance. The energy band structure of the three regions in equilibrium can again be derived easily by equalizing the Fermi levels, as shown in Figure 6.21(b).

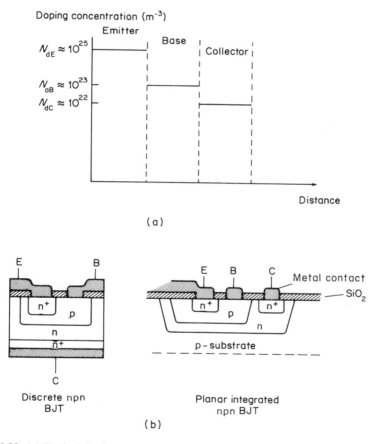

(a)

(b)

Figure 6.22. (a) Typical doping concentrations and (b) structures of npn bipolar junction transistors. (In practice the doping concentrations, especially in the emitter and base are much less uniform than implied by (a) above)

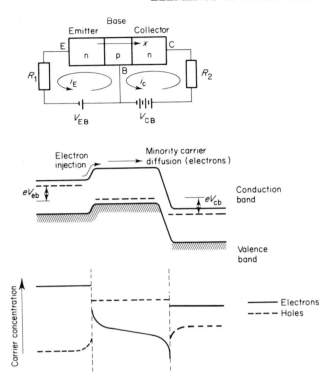

Figure 6.23. Common base operation of an npn transistor: (a) bias arrangements; (b) energy bands with biases applied; (c) resulting carrier concentrations

In both kinds of transistor the central region, which is called the *base*, is usually very narrow and rather lightly doped in comparison with the *emitter*, for reasons discussed below. The third region, the *collector*, usually has the lowest doping concentration. Typical doping concentrations are shown in Figure 6.22(a) while Figures 6.22(b) and 2.46(f) show some practical devices. The operation of junction transistors depends on the interaction of majority and minority carriers, and as both carrier types are involved such devices are classed as bipolar.

6.8.1   THE npn TRANSISTOR

Let us consider the effect of applied voltages on an npn transistor. With the arrangement shown in Figure 6.23, known as common base operation, we see that one of the pn junctions, the base–collector, is effectively reverse biased, while the other, the emitter–base, is forward biased.

The emitter region is so called since it *emits* charge carriers (in this case electrons); similarly the collector is so called since it *collects* charge carriers. For simplicity, the current flow will be assumed to take place only in the x direction. As the junction between the emitter and the base region is forward biased, electrons are 'injected' from the former into the latter. Very few holes are injected in the opposite direction because of the relatively low doping level in the base (more advanced discussions show that most efficient transistor action occurs if the current across this junction is almost exclusively carried by carriers originating in the emitter). The effect of the bias, therefore, is to maintain a certain excess electron concentration just inside the base region.

The base–collector junction is reverse biased, so that any electron minority carriers reaching the latter junction are swept across into the collector region. This has the effect of maintaining zero electron density just inside the base region at the base–collector junction. A minority carrier concentration gradient therefore exists across the base region (Figure 6.23c), and the electrons diffuse from the emitter–base junction towards the base–collector junction.

In practice the base of transistors, as mentioned earlier, is very narrow and lightly doped; these two factors ensure that virtually all the electrons injected into the base diffuse right across it without recombining with majority carriers. To further minimize recombination, it is also desirable to keep the number of defects or accidental impurities in the base region down to a minimum since these could act as minority carrier traps or recombination centres. Therefore, in practice, the current flowing in the collector circuit is only a very little less than that in the emitter circuit.

The base width is governed by the minority carrier lifetime τ_e, since it must not be larger than the distance covered by a carrier diffusing down the concentration gradient before it recombines. For a typical lifetime of $\approx 50 \, \mu s$ and a typical mobility of $\approx 0.15 \, m^2 \, V^{-1} \, s^{-1}$, a charge carrier travels about $0.4 \, mm$ before recombining. For maximum efficiency, it is usually recommended that the base should be not more than one-tenth of this value (that is $40 \, \mu m$).

6.8.2 THE pnp TRANSISTOR

The operation of a pnp transistor is exactly the same as for an npn transistor except, of course, that the roles of electrons and holes are interchanged (see Figure 6.24). Holes are injected from the emitter into the base region where, as minority carriers, they diffuse to the base–collector junction and are swept into the collector region, increasing the reverse bias saturation current of the base–collector pn junction. If the current in the emitter circuit is increased by altering the emitter–base potential so that more minority carriers are injected into the base, the reverse saturation current of the

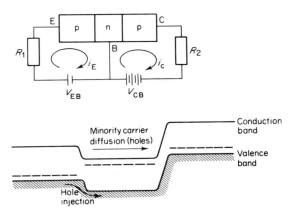

Figure 6.24. Common base operation of a pnp transistor

second junction increases, that is the current flowing in the collector circuit increases.

6.8.3 TRANSISTOR CHARACTERISTICS AND GAIN PARAMETERS

One of the main uses of the transistor is as an amplifying device. In the common base arrangement it has been seen that the collector current is almost but not quite equal to the emitter current. The common base current gain parameter α, which is defined as

$$\alpha = \frac{\text{change in collector current}}{\text{change in emitter current}} = \frac{\Delta I_C}{\Delta I_E} \qquad (6.20)$$

is then slightly less than unity, typically being between 0.9 and 0.999. The difference in the emitter and collector currents is accounted for by the base current.

The output characteristics of the transistor (Figure 6.25) show the dependence of collector current I_C on collector voltage V_{CB} for various values of emitter current I_E. It is seen that the collector current is almost independent of the collector voltage, providing that this is large enough to prevent majority carrier diffusion across the junction and ensure that saturation has occurred. As a consequence, the collector circuit has a high impedance (looking back into the transistor we see a reverse biased junction!). This is in contrast to the emitter circuit which has a low impedance, and small changes in emitter–base voltage produce large changes in emitter current; the emitter–base junction itself has a small impedance, being forward biased (see Section 6.7). Changes in emitter current, therefore, are determined by the input resistor R_1 so that

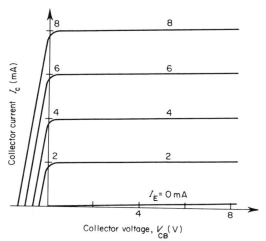

Figure 6.25. Transistor output characteristics for the common base operation of an npn transistor

$$\Delta I_E = \frac{\Delta V_{EB}}{R_1} \qquad (6.21)$$

The corresponding change in collector voltage is given by

$$\Delta V_{CB} = I_C R_2 \qquad (6.22)$$

where R_2, the output resistor, can be rather high; remember that the collector circuit has a high impedance! The voltage gain is then given by

$$\frac{\Delta V_{CB}}{\Delta V_{EB}} = \frac{\Delta I_C R_2}{\Delta I_E R_1} \approx \frac{R_2}{R_1} \qquad (6.23)$$

where R_2 and R_1 are typically $50\,\text{k}\Omega$ and $100\,\Omega$ respectively so that voltage gains ≈ 500 are reasonable.

This description of the operation of the device in terms of transferring almost constant current from a low-impedance to a high-impedance circuit gave the device its name, that is 'transfer resistor', subsequently shortened to 'transistor'. The input characteristic of the transistor is the normal forward biased junction curve for the emitter–base pn junction and corresponds to Figure 6.13.

Current gain can be obtained from the transistor in terms of the variation of the collector current with base current in the common emitter mode shown in Figure 6.26(a), which is a much more usual configuration. The corresponding characteristics are shown in Figure 6.26(b). The common emitter current gain parameter β is defined as

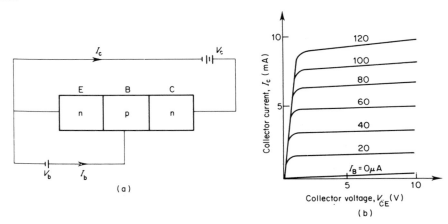

Figure 6.26. Common emitter operation of an npn transistor: (a) circuit configuration; (b) transistor output characteristics

$$\beta = \frac{\text{change in collector current}}{\text{change in base current}} = \frac{\Delta I_C}{\Delta I_B} \qquad (6.24)$$

Operation in this mode can be explained in terms of the requirement that the base should remain electrostatically neutral. Thus, suppose that a few additional holes flow into the base of an npn transistor as a result of a small increase in base current. These holes will result in many compensating additional electrons being injected from the emitter into the base. This is because there is an important difference in the times that the two carrier types remain in the base. The average excess hole remains in the base until it recombines, that is for a time approximately equal to the hole lifetime τ_h, which may be $\approx 10\ \mu s$. On the other hand, the injected electrons remain in the base only for the time τ_t that it takes them to diffuse across the base, which is typically $0.1\ \mu s$. Therefore, for each additional hole flowing into the base an additional τ_h/τ_t additional electrons can be injected into it and charge neutrality will be maintained.

Therefore the current gain is

$$\beta = \frac{\text{hole lifetime}}{\text{base transit time}} = \frac{\tau_h}{\tau_t} \approx 100$$

Not surprisingly, the gain parameters α and β are related. By Kirchoff's law we may write

$$I_B + I_E + I_C = 0$$

so that

$$\Delta I_B = -(\Delta I_C + \Delta I_E)$$

Substitution for ΔI_E from equation (6.20) then gives

$$\Delta I_B = -\Delta I_C\left(1 - \frac{1}{\alpha}\right)$$

Hence

$$\beta = \frac{\Delta I_C}{\Delta I_B} = \frac{\alpha}{1-\alpha} \qquad (6.25)$$

As α can be in the range 0.9–0.999, we see that β can vary from 10 to 1000.

In both modes of operation the transistor has a fairly low input impedance while, in many circuits, devices with high input impedances are required. In these cases field effect transistors (FET), which are described in Section 6.9, may be used.

6.8.4 SWITCHING MODE

The bipolar junction transistor is often used as an ON/OFF device in switching circuits; indeed this is the dominant use for the majority of integrated circuits. Let us consider the common emitter circuit shown in Figure 6.27(a), where the transistor is driven by a constant voltage V and a load resistor R_L is included in series with the collector. The relationship between the collector voltage V_{CE} and collector current I_C is given by

$$V = V_{CE} + I_C R_L \qquad (6.26)$$

Equation (6.26) represents the 'load line', which is shown plotted on the common emitter characteristics in Figure 6.27(b). When the transistor is used as a current amplifier, the base current I_B is chosen so that the transistor's operating point might typically be at the point Q.

Suppose now that the base current I_B is reduced to zero, so that both junctions are reverse biased; then I_C also becomes zero and V_{CE} becomes equal to the supply voltage V. This is the OFF state, which is represented by the point R in Figure 6.27(b); this is often referred to as the cut-off regime. If the base current is made fairly large and positive, the transistor switches from the cut-off regime to what is called the saturation regime. This corresponds to the ON state and is shown by point P in Figure 6.27(b). In the ON state the collector current is large, it is given by $I_C = V/R_L$, while the collector voltage falls to very small values. The bipolar junction transistor represents an almost ideal switch. It is capable of switching at very high speeds, typically $0.1\mu s$, and is therefore useful in fast data processing applications.

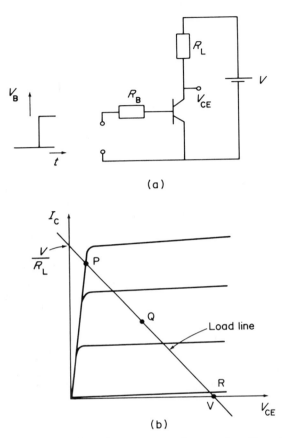

(a)

(b)

Figure 6.27. (a) An npn transistor in common emitter mode with a step voltage applied to the base, which switches the transistor from off to on. These points are represented by points R and P on the load line drawn superimposed on the common emitter characteristics shown in (b)

6.9   FIELD EFFECT TRANSISTORS

Field effect transistors (FETs), like bipolar junction transistors (BJTs), are used predominantly to amplify a small signal or to act as a switch in digital circuits. Although high-quality FETs were developed subsequently to high-quality BJTs they are often now used in preference, especially where a high input impedance is required. Field effect transistors have input impedances ranging from $10^5\ \Omega$ to perhaps $10^{12}\ \Omega$ or even higher.

Other advantages of the FET arise from the fundamental difference between FTEs and BJTs in that in the case of the former the current is carried predominantly by only the majority carriers (consequently the device is often referred to as unipolar). This results in the FET being rather insensitive to fluctuations in temperature, and to being less affected by other external factors such as nuclear radiation. Finally, the fact that a comparatively small area of chip is required to fabricate an FET gives them a high packing density, so they are attractive for use in integrated circuits.

Although there are many different types of FET, the basic operation results from the control of a current flowing along a semiconductor filament by the application of an electric field (hence the name). In fact the concept of a field effect transistor was first proposed by Lilienfield in 1930. The conductance G of the filament can be expressed as

$$G = \frac{1}{R} = \frac{ne\mu A}{L} \tag{6.27}$$

where A is the cross-sectional area of the filament and L its length. The applied electric field then varies either the effective (that is conducting) cross-sectional area A, or the carrier concentration n, or both of these parameters. There are two main categories, namely the junction field effect transistor (JFET) in which A varies and the insulated gate field effect transistor (IGFET), of which perhaps the most common is the metal oxide semiconductor transistor (MOST), in which both A and n vary.

Although the JFET has a relatively minor role in modern electronics we shall describe it first as its operation is rather easy to understand and it provides an introduction to the more technologically important MOST.

6.9.1 THE JUNCTION FIELD EFFECT TRANSISTOR

The JFET is shown schematically in Figure 6.28(a), while more practical geometries are shown in Figures 6.28(b) and 2.47. In essence the FET consists of a semiconducting filament called the *channel*, which here is n-type, with very heavily doped p^+-regions on either side. If these regions, the *gates*, which are electrically connected, are shorted to one end of the channel, the *source*, and a voltage V_d applied to the other end, the *drain*, then a current I_d will flow. This current will produce an IR drop along the channel and the potential at any point in it increases from the source, becoming more positive towards the drain. The n-type channel to p^+-gate pn junctions are therefore reverse-biased, becoming progressively more so towards the drain. In view of the relative doping levels of the channel and gates the junction depletion layer lies almost entirely within the channel. As the width of the depletion layers depends on the applied voltage they are wedge shaped, as shown in Figure 6.28. Since the depletion layers effectively behave as insulating regions, current flow is confined to the region between them.

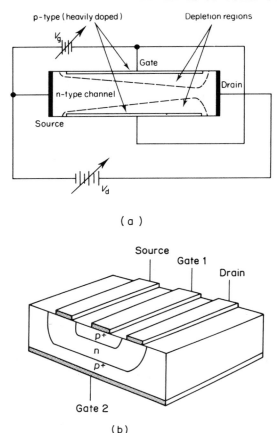

(a)

(b)

Figure 6.28. (a) Schematic representation of a junction field effect transistor showing the bias arrangement. (b) The planar form of the JFET suitable for inclusion in integrated circuits

As the drain voltage V_d increases the cross-sectional area of the conducting channel is steadily reduced so that, while initially I_d increases linearly with V_d, eventually this is no longer the case and I_d saturates at some value of $V_d = V_p$, called the 'pinch-off' voltage. At pinch-off it is imagined that the depletion layers meet, so closing off the conducting channel. This might imply that the drain current would become zero, but clearly this cannot be because then the potential drop causing the reverse bias would disappear. The actual mechanism of saturation is somewhat complicated and not fully understood. The I_d–V_d or drain characteristic for $V_g = 0$ is then as shown in Figure 6.29.

If now the gates are made negative, again reverse-biasing the gate to channel pn junction, then in the absence of drain current the depletion layers would

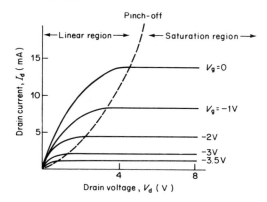

Figure 6.29. Typical current–voltage characteristics for a silicon n-channel JFET

extend uniformly into the channel, as shown in Figure 6.30. When drain current now flows then the wedge-shaped depletion layers are superimposed on the uniform ones so that the conducting region of the channel is greatly reduced. In this case pinch-off occurs at a lower value of drain voltage. Therefore the family of drain characteristics is as shown in Figure 6.29. It is seen that the reverse voltage applied to the gate modulates the drain current. In a typical application a voltage signal to be amplified is applied to the gate so that, in contrast to the BJT, we have a voltage-controlled amplifier. The signal is applied via a reverse biased pn junction so that, as mentioned above, the input impedance is very high.

6.9.2   THE METAL OXIDE SEMICONDUCTOR TRANSISTOR

In this type of FET the channel current is controlled by a voltage applied to a gate electrode which is isolated from the channel by an insulating layer of silicon dioxide. The structure is then as shown in Figure 6.31. Two heavily doped n-regions are diffused into a p-type substrate to form the source and drain. With no gate voltage applied there can be no current flow between the source and

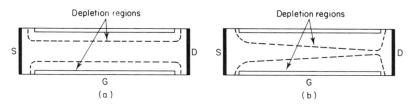

Figure 6.30. Diagrams showing the depletion layers in the channel of a JFET: (a) for $V_g > 0$ and $V_d \approx 0$; (b) for $V_g > 0$ and $V_d > 0$

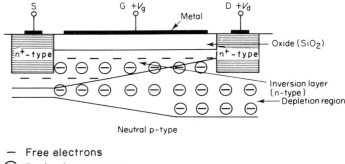

— Free electrons
⊖ Ionized acceptors

Figure 6.31. An induced-channel MOST with positive gate bias

drain because of the existence of back-to-back pn junctions, one of which will
be reverse biased whatever the polarity of the drain with respect to the source.

If, however, a positive voltage is applied to the gate an electric field is set
up across the oxide which terminates on negative charges induced in the upper
region of the p-type substrate. For small gate voltages the negative charges are
provided by holes being repelled from the upper region of the semiconductor,
leaving ionized acceptors and thereby creating a depletion layer. As explained
in Section 6.2 this causes the band edges of the semiconductor to bend
downwards. As the gate voltage is increased conduction electrons are drawn
into this region to provide the necessary additional negative charge so that
eventually, at the *threshold or turn-on voltage* V_T, an inversion layer is formed
in which the material has changed from p- to n-type. The energy bands have

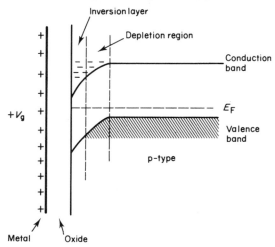

Figure 6.32. Energy band bending due to a positive voltage applied to the gate of a MOST

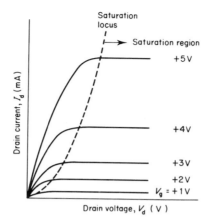

Figure 6.33. Current–voltage characteristics of an induced n-channel MOST

bent down to such an extent that the Fermi level has become nearer the conduction band than the valence band, as shown in Figure 6.32. The inversion layer acts as a conducting channel between the source and drain; obviously its width is governed by the magnitude of the gate voltage V_g. Increasing V_g increases the channel width and therefore the drain current for a given value of drain voltage. The drain characteristics are as shown in Figure 6.33. This is often referred to as an enhancement mode of operation as opposed to the depletion mode of the JFET.

If there is an applied drain voltage in addition to the gate voltage then a wedge-shaped depletion layer will form. As with the JFET, pinch-off will occur, for a given value of gate voltage, when the reverse bias between the induced n-channel and p-type substrate is such that the effective cross-sectional area of the conduction channel becomes zero at the drain.

The MOST has an extremely high input impedance as the signal to be amplified is applied to the gate, which is separated from the semiconductor by the layer of insulating silicon dioxide, which may have a resistance of up to 10^{14}–$10^{15} \, \Omega$. As with metal–semiconductor junctions the operation of the MOST is complicated by the presence of surface states. In this case there is usually a layer of fixed surface charge at the semiconductor–oxide interface. The control of this surface charge was one of the factors which delayed reproducible manufacture of the MOST.

Apart from this device, which is often called the induced-channel MOST, there are MOSTs in which a conducting n-channel between the source and drain is diffused into the p-substrate during the fabrication process, as shown in Figures 6.34 and 2.48. A little thought will lead us to realize that we can either increase the width of this conducting channel by applying a positive gate voltage or decrease its width by applying a negative gate voltage. We see therefore that the device can operate in both enhancement and depletion modes. Diffused-channel

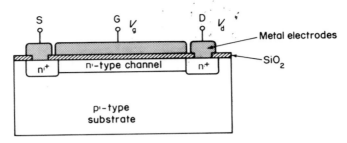

Figure 6.34. Diffused channel MOST

MOSTs are often referred to being 'normally on' as there is a conducting channel, even in the absence of a gate voltage. As already mentioned, the MOST is particularly useful in digital integrated circuits, in which it can easily and quickly be switched from an OFF state (no conducting channel) to an ON state.

Both n- and p-channel FTEs are available. As with BJTs, however, advantage is often taken of the fact that the electron mobility in silicon is much higher than the hole mobility, so that n-channel devices are generally preferred.

6.10   OPTOELECTRONIC DEVICES

In parallel with the progress in solid state physics, in the last two decades there has been a significant development in optical devices which has evolved into what we call optoelectronics. Thus, for example, there has been a marked increase in the variety of lasers and their applications, the almost incredible rate of development and installation of fibre optic communication systems and increasing interest in new display devices and optically based computers. All of these branches of optoelectronics depend on, or include, semiconductor devices such as light emitters and detectors. The operation of the majority of these depends on phenomena occurring at pn junctions.

6.10.1   PHOTODETECTORS

Photoconductors

It was seen in Section 5.5 that if photons of energy $h\nu$ are incident on a semiconductor, whose energy gap $E_g < h\nu$, then additional electron–hole pairs will be generated. The number of additional carriers will be in proportion to the irradiance $\mathcal{I}$ (in watts per square metre) of the incident light, that is on the number of photons falling on, and absorbed by, the semiconductor. Thus, if it is supposed that the concentrations of excess carriers are Δn and Δp, where $\Delta n = \Delta p$, there will be an increase in the conductivity of the semiconductor given by (see equation 5.2)

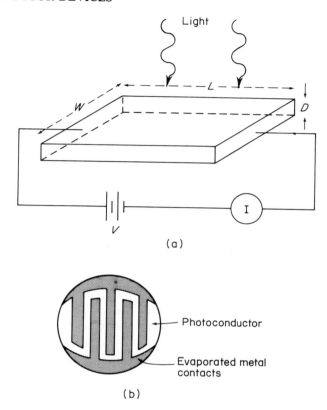

Figure 6.35. (a) Slab of photoconductive material of length L, width W and thickness D. Opposite faces have electrodes attached and a voltage V is applied as shown. The radiation falls on the upper surface. (b) Top view of a typical photoconductive cell. The comb-like structure ensures maximum sensitivity

$$\Delta\sigma = \Delta ne\mu_e + \Delta pe\mu_h$$

or

$$\Delta\sigma = \Delta ne(\mu_e + \mu_h) \tag{6.28}$$

If a voltage V is applied across a slab of semiconductor of dimensions W,L,D, as shown in Figure 6.35(a), the increase in current, that is the photocurrent, resulting from the presence of the excess carriers will be

$$\Delta i = WD\Delta ne(\mu_e + \mu_h)\frac{V}{L} \tag{6.29}$$

The excess carrier concentration Δn depends on the balance between the rate of optical generation, that is on the irradiance, and the rate of recombination of the carriers. It can be shown that

$$\Delta n = \Delta p = \frac{\eta \mathcal{I} \tau_c}{h\nu D} \qquad (6.30)$$

where τ_c is the carrier lifetime and η is the quantum efficiency, which expresses the probability that an incident photon will indeed generate an electron–hole pair. Therefore, substituting equation (6.30) into equation (6.29) gives

$$\Delta i = \frac{W\eta \mathcal{I} \tau_c e}{h\nu L} \; (\mu_e + \mu_h) V \qquad (6.31)$$

Hence the photocurrent is directly proportional to the irradiance. It can be seen that, for a given irradiance, the photocurrent increases if W is increased and L decreased, and consequently the photoconductive material is often fabricated in the comb-like structure of Figure 6.35(b).

A wide range of photoconductive materials is available so that the wavelength range from about 350 nm to 30 μm can be covered. These include cadmium sulphide (CdS), which is widely used in the visible spectrum for camera lightmeters, lead sulphide (PbS), indium antimonide (InSb) and mercury cadmium telluride (HgCdTe). One of the main disadvantages of photoconductive detectors is their poor response times, typically of the order of milliseconds, whereas for many applications fast responses in the nanosecond range are required.

6.10.2   PHOTODIODES

If light of frequency $\nu \geqslant E_g/h$ is incident on the depletion layer of a pn junction and the bulk regions immediately adjacent to it, the additional electrons and holes so created may diffuse to the junction and be separated by the electric field caused by the contact potential difference V_0, as shown in Figure 6.36(a). In particular, holes will be swept into the p-region and electrons into the n-region, giving rise to an additional current I_{op} directed from the n- to the p-region. This is in the opposite sense to conventional forward current flow so that the diode equation (6.16) becomes

$$I = I_0 [\exp(hV/kT) - 1] - I_{op} \qquad (6.32)$$

and the I–V curves are lowered by an amount I_{op}, which depends principally on the irradiance of the illumination (Figure 6.36b).

If the irradiance is $\mathcal{I}$, the sensitive area of device near the junction is A and the quantum efficiency is η; then the number of electron–hole pairs generated per second is

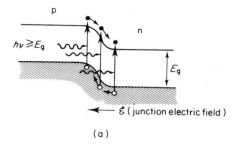

(a)

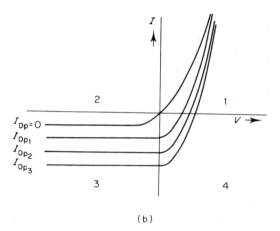

(b)

Figure 6.36. Absorption of optical radiation near a pn junction: (a) creation of additional carriers near the junction; (b) the effect of the radiation on the I–V characteristic (in quadrant 3 the device is used as a photodiode, in quadrant 4 as a solar cell)

$$N = \frac{\mathcal{I}A\eta}{h\nu} = \frac{\mathcal{I}A\eta\lambda}{hc}$$

assuming that the photons all have the same energy $h\nu$. The resultant current I_{op} is then simply Ne or

$$I_{op} = \frac{eA\mathcal{I}\eta\lambda}{hc} \tag{6.33}$$

If the device is now short circuited, that is $V = 0$ in equation (6.32), the diffusion and drift currents of course cancel and the current flow becomes $I = -I_{op}$ so that the I–V characteristics cut the I axis at negative values directly proportional to the irradiance of the incident light. This is referred to as the photoconductive mode of operation.

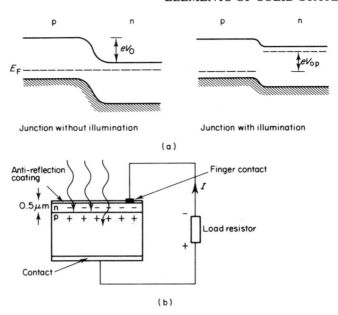

Figure 6.37. The solar cell: (a) the effects of illumination on the energy bands; (b) typical structure of a solar cell

If, on the other hand, the illuminated pn junction is open circuited then the current I is zero and a voltage V_{op} develops across the junction, where from equation (6.32) we have

$$V_{op} = \frac{kT}{e} \ln\left(1 + \frac{I_{op}}{I_0}\right) \tag{6.34}$$

In this case, the photovoltaic mode of operation, the output voltage, is proportional to the logarithm of the irradiance. Despite the form of equation (6.34), V_{op} does not increase indefinitely as the irradiance increases. In fact, the maximum value of V_{op} is approximately the contact potential difference V_0 (which in turn approximates E_g/e), as can be seen from Figure 6.37(a), and in silicon, for example, is about 0.6 V.

The spectral response of such photodiodes, as with photoconductive detectors, depends on the energy gap of the semiconductor used to form the pn junction. The spectral range from about 350 nm to 100 μm can be covered using diodes fabricated from materials such as zinc sulphide, gallium arsenide, silicon and germanium doped with zinc or boron. Unfortunately the latter have to be operated at temperatures as low as 4 K to reduce noise due to thermally generated carriers.

The optically generated voltage V_{op} and current I_{op} provide the source of power in solar cells, which are finding increasing applications, ranging from power sources for third world villages to communications satellites. In the fourth quadrant of the I–V characteristics we note that while the voltage is positive, the current is negative, so that the power dissipated in the device is negative, that is the device produces power in the external circuit.

Solar cells as shown in Figure 6.37(b) have a large area junction located near to the surface of the device, so that the majority of carriers generated by the incident photons can diffuse to the junction before recombining. The surface of the device is usually contacted with a very fine metal grid and coated with an antireflection coating to maximize the efficiency. The efficiency of single crystal silicon cells, in terms of converting optical to electrical energy, is typically 15%, while devices made from amorphous silicon have efficiencies of 8–10%. There is currently a great deal of interest in cells based on GaAs and indium sulphide InS for space applications. Alternatively, thin film solar cells can be formed by evaporating thin film junctions of materials such as CdS to give large areas at low costs.

6.10.3 LIGHT EMISSION

In Section 5.5.1 it was seen that direct recombination of electrons in the conduction band with holes in the valence band in semiconductors such as GaAs is usually accompanied by the emission of a photon. Similarly, recombination involving donor and acceptor impurities may result in radiation being emitted. These recombination processes are the basis of light emission by light-emitting diodes and laser diodes. It was also seen in Chapter 1 that when an electron falls from an excited state to a lower energy state a photon is emitted; this process is the basis of many other lasers and light sources.

Light-emitting diodes

In forward biased junctions there can be appreciable recombination of the injected carriers in and close to the depletion region (Figure 6.38a). In those semiconductors, such as GaAs, where direct radiative recombination is likely, recombination in the depletion layer results in the emission of photons of energy $h\nu \approx E_g$ (this is injection electroluminescence). These light-emitting diodes (LEDs) have many uses as display and warning devices.

In GaAs LEDs, where the energy gap $E_g = 1.43$ eV, the wavelength of the peak of the emitted radiation is 890 nm, which is in the infrared. However, by using mixed crystals of gallium arsenide and gallium phosphide ($E_g = 2.26$ eV), that is $GaAs_{1-x}P_x$, the energy gap can be increased and radiation in the visible part of the spectrum can be produced. The limit to this is when the fraction x of phosphorus exceeds 45%, at which point radiative recombination becomes very unlikely as the crystal changes from having a direct to an indirect energy gap. The wavelength range can be further increased by suitable doping of GaP;

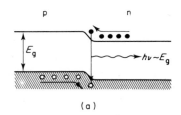

(a)

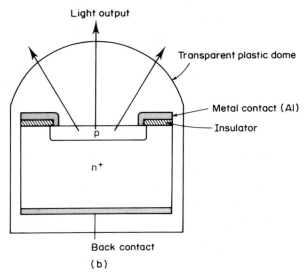

(b)

Figure 6.38. The light emitting diode: (a) recombination of electrons and holes at a forward biased junction to produce photons; (b) typical structure of an encapsulated LED

for example GaP diodes doped with sulphur and zinc emit in the green while when doped with zinc and oxygen they emit in the red. Recently LEDs emitting blue light based on zinc sulphide (ZnS) and silicon carbide (SiC) have been developed.

In LEDs the junction should again be shallow to minimize reabsorption of the emitted light. Also to avoid internal reflection of the light the junction is often encapsulated in a hemisphere of clear glass or plastic, as shown in Figure 6.38(b). Despite these innovations the external quantum efficiency (a measure of the optical output relative to the electrical input) is only a few per cent, even though the internal quantum efficiency (a measure of the photons generated compared to the carriers injected into the depletion layer) can be very high, perhaps 80–90%.

Semiconductor diode lasers—fundamentals of laser operation

The light emitted by an LED, and indeed most other sources of light, is *incoherent*, by which is meant that there is no ordered phase relationships between the electromagnetic waves emitted. The light also has quite a wide spectral range, perhaps 30 to 60 nm depending on the basic material used. In contrast, lasers emit light that is coherent and highly monochromatic.

If an electron is excited from its ground state energy E_1 to an excited state E_2, it will normally remain in that state for only a very short time before *spontaneously* falling back to the ground state (Figure 6.39a), with the emission of a photon of energy $h\nu = E_2 - E_1$. If there are many such excited electrons they will emit photons at random times and in random directions and so generate incoherent light.

If, on the other hand, a photon of energy $E_2 - E_1$ passes near an electron in the energy state E_1 it will be *absorbed* and the electron will be excited to the state E_2 (Figure 6.39b). Einstein, in 1917, recognized the possibility

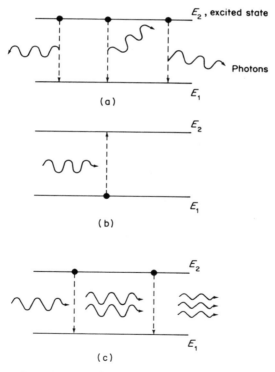

Figure 6.39. Energy level diagram illustrating: (a) spontaneous emission, the photons are emitted randomly; (b) absorption; (c) stimulated emission, the emitted photons are coherent with the stimulating photon

of a third process involving the interaction of electrons and photons, namely *stimulated emission*. In this process a photon of energy $E_2 - E_1$ may interact with an electron in the excited state E_2, causing it to emit a photon (Figure 6.39c). There is nothing too surprising about this basic process, as the electron in state E_2 would, sooner or later, return to the state E_1 with the emission of a photon anyway. What is surprising is the fact that the electromagnetic wave associated with the stimulated photon is coherent with that associated with the stimulating photon; that is it has the same phase, state of polarization and direction of travel as well as the same frequency. Therefore, if an optical signal is passed into a medium in which there are many electrons in the excited state E_2, the signal may be amplified; that is we have Light Amplification by Stimulated Emission of Radiation.

In general, the medium will be in a condition of thermal equilibrium in which there are many, many more electrons in the ground state than in the excited state so that the incident optical signal will be absorbed. To avoid this and achieve amplification the equilibrium state must be destroyed and what is called a *population inversion* created. In this condition there are many more electrons in the excited state than in the ground state. To create a population inversion energy must be *pumped* into the medium by some process such as the passage of a current, the creation of an electrical discharge or illumination with optical radiation.

In the vast majority of lasers, the laser medium or optical amplifier is placed between two parallel, highly reflecting mirrors. The generation of a coherent light beam can now take place without the need for an incident signal wave. Photons produced by spontaneous emission are reflected to and fro between the mirrors, building up an intense coherent beam by stimulated emission. One of the mirrors is usually made a little less reflective than the other so that some of this light can escape as the useful output.

The two mirrors form a *resonant cavity* and provide optical feedback so that the laser in reality is an optical oscillator. The parallel with an electronic amplifier in which positive feedback is used to create an oscillator is obvious. Laser oscillations will become established for wavelengths which satisfy the condition

$$p\lambda = 2L \qquad (6.35)$$

where p is an integer and L is the length of the resonant cavity, providing of course that there is energy available at the particular wavelength and that there is sufficient optical gain to counterbalance the inevitable losses.

Semiconductor pn junction lasers are not very different in principle from the LEDs discussed in the previous section. In addition to the conditions for light emission, however, a population inversion must be created and a resonant cavity provided. A population inversion can be created in the depletion layer of a junction fabricated from degenerate p- and n-type semiconductors by applying

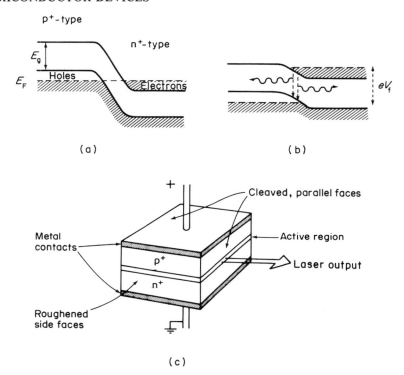

Figure 6.40. The semiconductor pn junction laser: (a) shows the energy bands in equilibrium and (b) with a forward bias V_f ($\approx E_g/e$) applied; (c) shows a simple, homojunction structure

a large forward bias, that is a bias $\approx E_g/e$ volts. The doping concentrations in junction lasers are even higher than those in the tunnel diode discussed in Section 6.5. The energy bands of the junction in equilibrium and with a forward bias applied are then as shown in Figure 6.40(a). It can be seen that in effect the forward bias has created a situation in the depletion layer in which there are many free electrons in excited states with respect to unfilled lower energy states (holes), that is a population inversion has been formed.

The resonant cavity in this case is usually provided by a pair of parallel cleaved crystal faces. The reflectance of these faces due to Fresnel reflection, $R = [(n_1 - n_2)/(n_1 + n_2)]^2$, is just large enough to enable laser action to occur. The crystal faces in the perpendicular direction are usually deliberately roughened to prevent laser action in that direction also.

Pumping of the laser medium is then achieved by the passage of a large current through the forward biased junction. At lowish values of the current the device behaves as an LED, but as the current is increased a threshold is reached at which laser action commences and there is then

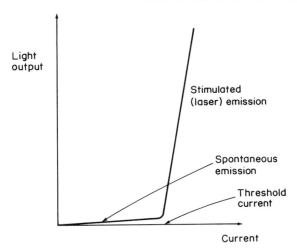

Figure 6.41. Light output–current characteristic of an ideal semiconductor pn junction laser

a rapid increase in the optical output with a further small increase in current, as shown in Figure 6.41.

In the case of the homojunction lasers of the type described here current densities of some $400 \, \text{A mm}^{-2}$ are required to reach threshold. As the dimensions of a laser diode are typically $300–400 \, \mu\text{m}$ it is easy to realize that current densities of this magnitude will cause considerable heating within the device. Consequently, homojunction lasers can only be operated in very short pulses and even then they must be cooled to low temperatures.

Recently considerable improvements in performance have been achieved by the use of heterojunctions, in which there are different semiconductors on the two sides of the junction. Heterojunction lasers, in which there are perhaps five parallel layers of semiconductors such as GaAs and GaAlAs (gallium aluminium arsenide), have threshold current densities of only a few hundred amperes per square millimetre, corresponding to actual currents of only $20–30 \, \text{mA}$. These devices can operate continuously, even at room temperature, and deliver optical powers of over $10 \, \text{mW}$. Such devices are widely used in compact disc equipment to 'read' the stored information.

Doped insulator lasers

While strictly speaking the so-called doped insulator lasers are not semiconducting devices, they do involve insulating crystals, which are deliberately doped with impurities to change their characteristics. As an example, let us consider the Nd:YAG laser, which comprises a single crystal of YAG (yttrium aluminium garnet, $Y_3Al_5O_{12}$) which incorporates up to one per cent of Nd^{3+}

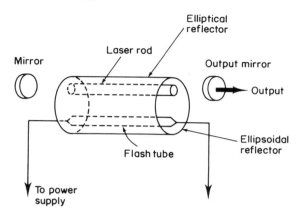

Figure 6.42. Typical construction of a doped insulator laser such as Nd : YAG

(that is triply ionized neodymium atoms), replacing yttrium Y^{3+} ions in the crystal lattice. The neodymium ions provide a set of energy levels in which a population inversion can be created by exposing the crystal to an intense flash of white light, or indeed to light from a pn junction laser with a suitable wavelength output. Electrons in the Nd^{3+} ions absorb the incident light and are excited to appropriate high energy levels. Stimulated emission can then occur. The YAG crystal, usually referred to as the host, has a secondary, but crucial, role in terms of modifying the energy levels of the Nd^{3+} ions; we shall not pursue this further.

In a typical YAG laser, shown schematically in Figure 6.42, light from a flash tube is concentrated on to the crystal by arranging for the laser rod (perhaps 100 mm long by 10 mm in diameter) and the flash tube to lie along the focii of an elliptical reflector. The resonant cavity is formed either by polishing and coating the ends of the laser rod with a reflective coating or by using external mirrors.

As the pumping is provided by a short flash of white light lasting ≈ 1 ms, the laser output will be in the form of pulses. A continuous output can be obtained by using a junction laser to excite the electrons. In this case the incident photons have exactly the right energy to create the population inversion and pumping is very much more efficient than with a white light source, in which most of the radiation is ineffective and simply creates unwanted, excess heat.

In pulse mode Nd:YAG lasers can produce extremely high power pulses, which are widely used in a variety of metal forming processes such as cutting, drilling and welding and in medicine for general surgery and eye surgery.

6.10 CONCLUDING REMARKS

In this book we have provided a description of many of the important properties of solids at a level which will serve as an introduction to more advanced texts and research papers. Similarly we have described the operation of a selection of devices, particularly semiconducting devices, so that the reader will be able to understand more detailed descriptions of these, and related devices, and also the operation of the new devices that will inevitably be introduced.

During the last two to three decades integrated circuits have become more and more sophisticated so that we have moved through the periods of large scale integration (LSI), very large scale integration (VLSI) to ultra large scale integration (ULSI) in which device geometries are submicrometre in size and over ten million components may be present in a single chip. What of the future? It seems inevitable that devices based on silicon will continue to dominate, despite recent, increased interest in devices based on gallium arsenide, which because of the high electron mobility are useful at very high frequencies. Gallium arsenide technology has also developed in relation to optoelectronic devices, including the recently introduced quantum well devices. In these, which comprise, for example, alternate layers of gallium arsenide and aluminium arsenide some $10\,\mu$m thick, the energy gap and hence the wavelength of emission of light-emitting devices can be 'engineered' by varying the layer thickness. These techniques, and indeed the increasing integration of optical and electronic devices, will undoubtedly become more widespread and more important. Optical computers are but one area of potential widespread application.

Another recent development has been that of organic semiconductors, leading to what is sometimes called molecular electronics. There is interest in this area in relation to the development of chemical and biosensors. Interestingly much of this work is based on Langmuir–Blodgett films which were developed by I. Langmuir and C. Blodgett in 1930.

Solid state circuitry has made enormous contributions to medicine, to communications, to commerce and to manufacturing processes, and it is to be expected that improved manufacturing techniques and the availability of even cheaper and more reliable integrated circuits will allow a whole range of new products and applications to be considered.

Solid state physics is in a state of continual flux and one should never be reluctant to try out new ideas, even when they apparently contradict well-established theories. After all, the subject had its origins in the apparent contradiction of the known laws of classical physics, and it is an essential part of the development of any branch of science that ideas should be continually revised and modified in the light of new experimental facts.

PROBLEMS

Use the data provided in Appendix 1 as required.

6.1 Draw the energy bands for metal to p-type semiconductor junctions when the work function of the metal is (a) greater and (b) smaller than the work function of the semiconductor.

6.2 An abrupt silicon pn junction diode is formed from p- and n-material having resistivities of $1.3 \times 10^{-4}\,\Omega\,m$ and $4.62 \times 10^{-2}\,\Omega\,m$ respectively. Calculate the equilibrium contact potential and the depletion layer width, assuming that the temperature is 300 K. Also calculate the proportions of the depletion layer which are in the p- and n-regions respectively.

6.3 The current flowing in a particular pn junction at room temperature is $5 \times 10^{-12}\,A$ when a large reverse bias voltage is applied. Calculate the current flowing for forward biases of 0.4 and 0.6 V. Also calculate the forward slope resistance in each case.

6.4 Calculate the change in junction capacitance of the diode in Problem 6.2 as the bias voltage is increased from zero to $-5\,V$, given that the area of the junction is $5 \times 10^{-7}\,m^2$.

6.5 (a) Assuming that the electric field across the depletion layer of a pn junction is uniform (it isn't!) show that the breakdown voltage for avalanche breakdown can be written as

$$V_{br} = \frac{2E_g^2 \epsilon_0 \epsilon_r}{e^3 \lambda^2}\left(\frac{1}{N_a} + \frac{1}{N_d}\right)$$

where λ is the mean free path of the electrons in the depletion layer. (*Hint*: write down the condition that the energy gained by electrons between collisions should equal the energy gap of the semiconductor so that the electron will ionize any atom with which it collides.)

(b) Calculate the avalanche breakdown voltage of the diode specified in Problem 6.2, given that the mean free path $\lambda = 2.0 \times 10^{-7}\,m$.

6.6 Given that the time taken for charge carriers to diffuse a distance d is $d^2/2D$, where D is the diffusion coefficient, calculate the gain parameters α and β for a silicon npn transistor with a base width of $4\,\mu m$. You may assume that the carrier lifetime in the base is $1\,\mu s$ and that the temperature is 300 K.

6.7 Show that the pinch-off voltage V_p of a JFET can be written as

$$V_p = \frac{eN_d a^2}{2\epsilon_0 \epsilon_r} - V_0$$

where $2a$ is the width of the channel. (*Hint*: assume that the depletion layers touch one another at pinch-off.)

The resistivity of the n-type channel of a JFET is $0.023\,\Omega\,\text{m}$ and the channel width is $4\,\mu\text{m}$. Estimate the pinch-off voltage if the contact potential difference $V_0 = 0.7\,\text{V}$.

6.8 Choose materials suitable for the manufacture of the following devices
   (a) a pn junction photodiode sensitive to wavelengths less than (i) $1\,\mu\text{m}$ and (ii) $550\,\text{nm}$,
   (b) a LED emitting at a wavelength of approximately (i) $870\,\text{nm}$ and (ii) $345\,\text{nm}$.

6.9 Calculate the photoconductive current flowing through a piece of gallium arsenide which is 20 mm wide by 5 mm long subjected to very bright sunlight that has an irradiance of $1\,\text{kW}\,\text{m}^{-2}$. A voltage of 5 V is applied across the 5 mm length, the carrier lifetime is $7 \times 10^{-6}\,\text{s}$ and the quantum efficiency is 0.8; assume that the average frequency of the light that can generate photocarriers is $4.0 \times 10^{14}\,\text{Hz}$.

6.10 A large area photodiode has an area of $2 \times 10^{-4}\,\text{m}^2$. Calculate the current produced by light of irradiance $100\,\text{W}\,\text{m}^{-2}$ with an average wavelength of $0.7\,\mu\text{m}$; assume that the quantum efficiency is unity. If the reverse bias saturation current of the junction is $10^{-8}\,\text{A}$ calculate the voltage generated if the device is used in the photovoltaic mode.

6.11 If $1.0\,\text{nA}$ is the minimum current flowing through a photodiode which can be measured, calculate the sensitivity of the photodiode. Assume that the area of the junction is $25 \times 10^{-6}\,\text{mm}^2$, the quantum efficiency is 0.8 and the average wavelength is $0.7\,\mu\text{m}$. Also calculate how many photons are falling on the photodiode per second and the responsibility of the device in amperes per watt.

Suggested Reading List

Books marked with an asterisk are suitable for introductory reading.

GENERAL SOLID STATE THEORY

Blakemore, J. S., *Solid State Physics*, 2nd ed., Cambridge University Press (1985).

Coles, B. R. and Caplin, A. D., *The Electronic Structure of Solids*, Edward Arnold (1976).

Davies, D. A., *Waves, Atoms and Solids*, Longman (1978).

Fraser, D. A., *The Physics of Semiconductor Devices*, Clarendon Press (1984).

Kittel, C., *Introduction to Solid State Physics*, 6th ed., John Wiley (1986).

*Myers, H. P., *Introductory Solid State Physics*, Taylor and Francis (1990).

Rosenberg, H. M., *The Solid State*, 3rd ed., Oxford University Press (1986).

*Solymar, L. and Walsh, D., *Lectures on the Electrical Properties of Materials*, 4th ed., Oxford University Press (1988).

Scientific American, September 1967, Special Edition on the Properties of Materials.

*Tuck, B. and Christopoulos, C., *Physical Electronics*, Edward Arnold (1986).

CHAPTER 1

*Beiser, A., *Perspectives of Modern Physics*, McGraw-Hill (1969).

*Bueche, F. J., *Introduction to Physics for Scientists and Engineers*, 4th ed., McGraw-Hill (1986).

Semat, H. and Albright, J. R., *Introduction to Atomic and Nuclear Physics*, Holt, Rinehart and Winston (1972).

CHAPTER 2

*Allison, J., *Electronic Engineering Semiconductors and Devices*, 2nd ed., McGraw-Hill (1989).

Azaroff, L. V., *Elements of X-ray Crystallography*, McGraw-Hill (1968).

*Henderson, B., *Defects in Crystalline Solids*, Arnold (1972).

Holden, A., *Bonds Between Atoms*, Oxford (1971).

*Morgan, D. V. and Board, K., *An Introduction to Semiconductor Micro-technology*, John Wiley (1983).
Pamplin, B. R., *Crystal Growth*, 2nd ed., Pergamon Press (1980).
Scientific American, 1977, Special Edition on Microelectronics.

CHAPTER 3

*Bitter, F., *Magnets*, Heinemann (1960).
Carey, R. and Isaac, E. D., *Magnetic Domains and Techniques for Their Observation*, English Universities Press (1966).
*Grieg, D., *Electrons in Metals and Semiconductors*, McGraw-Hill (1969).
*Inokuchi, H. J., *Electrical Conduction in Solids*, Routledge and Keegan Paul (1965).
*Martin, D. H., *Magnetism in Solids*, Iliffe (1967).
Nussbaum, A., *Electronic and Magnetic Properties of Materials*, Prentice-Hall (1967).

CHAPTER 4

Blakemore, J. S., *Solid State Physics*, 2nd ed., Cambridge University Press (1985).
Davies, D. A., *Waves, Atoms and Solids*, Longman (1978).
Grieg, D., *Electrons in Metals and Semiconductors*, McGraw-Hill (1969).
*Heitler, W., *Elementary Wave Mechanics*, Oxford (1956).
Kittel, C., *Quantum Theory of Solids*, 2nd ed., John Wiley (1987).
Rosenberg, H. M., *The Solid State*, 3rd ed., Oxford University Press (1980).
Solymar, L. and Walsh, D., *Lectures on the Electrical Properties of Materials*, 4th ed., Oxford University Press (1988).

CHAPTER 5

See references for Chapter 4.

*Allison, J., *Electronic Engineering Semiconductors and Devices*, 2nd ed., McGraw-Hill (1989).
Smith, R. A., *Semiconductors*, 2nd ed., Cambridge University Press (1979).

CHAPTER 6

*Allison, J., *Electronic Engineering Semiconductors and Devices*, 2nd ed., McGraw-Hill (1989).

Connor, F. R., *Electronic Devices*, Edward Arnold (1980).

*Cooke, M. J., *Semiconductor Devices*, Prentice-Hall (1990).

Millman, J., *Microelectronics, Digital and Analog Circuits and Systems*, 2nd ed., McGraw-Hill (1987).

*Olsen, G. H., *Electronics, A Course Book for Students*, 2nd ed., Butterworths (1982).

*Olsen, G. H., *Electronics Made Simple*, Butterworths (1992).

Streetman, B. G., *Solid State Electronic Devices*, 3rd ed., Prentice-Hall (1988).

Sze, S. M., *Physics of Semiconductor Devices*, 2nd ed., John Wiley (1981).

*Wilson, J. and Hawkes, J. F. B., *Optoelectronics, An Introduction*, 2nd ed., Prentice-Hall (1989).

Discussion Questions

Describe the photoelectric effect and electron diffraction and discuss their significance in the development of modern physics.

Explain what is meant by the *uncertainty principle* and the *Pauli exclusion principle*.

What is meant by the electron configuration of an atom? Describe how you would determine the electron configuration of silicon (atomic number 14) and of iron (atomic number 26). Why is the configuration of iron not quite what you might expect?

What were Bohr's postulates? Why do we often describe Bohr's theory as being 'semi-classical'?

List some of the changes that might occur if Planck's constant were suddenly to become large and all the other laws of physics remain the same.

Offer an explanation as to why the fifth Balmer series line was not discovered along with the other four.

Is the gravitational attraction between the proton and the electron in the hydrogen atom significant in comparison with the Coulombic interaction? How would their ratio vary from orbit to orbit?

Why must hydrogen be at a fairly high temperature before emitting light?

Discuss the possible results that might be observed if the Tolman and Stewart experiment were carried out with a superconducting wire wound round the edge of the rotating wheel.

What were the assumptions made by Drude in his theory of the electrical conductivity of metals? What were the difficulties encountered by this theory?

Define the terms (a) thermal velocity, (b) drift velocity and (c) mobility.

Distinguish between drift and diffusion currents in semiconductors. Explain how electron and hole drift currents are 'additive' while electron and hole diffusion currents are 'subtractive'.

State in words the *Wiedemann–Franz law*. In view of the breakdown of the classical theory of conduction, why is it that the Wiedemann–Franz law can be established correctly using classical theory?

Discuss how the addition of small amounts of nickel to otherwise pure copper affects the way in which the conductivity varies with temperature.

Discuss the significance of the exchange integral in magnetic phenomena and indicate how its magnitude and sign govern the magnetic state of a given material.

Explain the form of the hysteresis loop for a ferromagnetic material in terms of the *domain theory*.

Discuss the significance of positive Hall coefficients.

The eigenvalues for a particle in a three-dimensional box are given by

$$E = \frac{h^2}{8mL^2} \, (n_1^2 + n_2^2 + n_3^2)$$

Use this expression to discuss degeneracy in electronic energy levels.

Give a physical interpretation of the wave function ψ which is a solution of Schrödinger's equation. What value would you expect ψ to have in the forbidden energy gap of an intrinsic semiconductor?

Explain the term *degeneracy* and describe how its removal leads to the formation of energy bands in solids.

Discuss the statement: 'When we know the eigenvalues the problem is solved and its correctness checked.'

Sketch the classical and quantum mechanical energy distribution curves and describe how the latter is used to account for the specific heat capacity of metals only being about 25.7 J mol^{-1} K^{-1}.

Use the uncertainty principle to describe the electron distributions for a particle in a three-dimensional box.

According to the E–k curves the effective mass of an electron becomes infinite at $k = \lambda/2a$. Give an explanation of this.

Why are semiconductors opaque in the visible, yet transparent in the near infrared?

Describe how the dynamics of minority carriers in a semiconductor can be investigated.

Explain carefully what is meant by the Fermi level and draw sketches to show how the electrical properties of various materials can be inferred from the positions of the Fermi level in each case.

What is meant by a positive hole? Explain how holes may be generated.

Describe two ways in which the energy gap of a semiconductor may be measured.

Explain why the conductivity of a metal decreases with increasing temperature whereas that of a semiconductor in general increases.

Describe how an intrinsic semiconductor can be made extrinsic by the addition of small amounts of suitable impurity.

Discuss thermionic emission and mention three or four situations where thermionic emission is important.

Each letter of the Roman alphabet belongs to one of the following five symmetry classes. List the letters in their appropriate classes:

1. No symmetry
2. Twofold symmetry axis
3. Vertical symmetry plane
4. Horizontal symmetry plane
5. All of 2, 3 and 4

If you know other alphabets, for example the Greek and the Russian, repeat the above exercise for those. Comment!

To what extent is it true to say that a crystal lattice is virtually open space?

What is a covalent bond? Explain how covalent bonds are arranged in the diamond structure. How do the diamond and zinc blende structures differ?

Describe the zone refining technique for the production of pure seed crystals. Outline the methods whereby such pure material can be converted into large single crystals. What is the largest diameter of silicon crystals used for the manufacture of integrated circuits?

Outline the procedure for finding the Miller indices of a given crystallographic plane and sketch the (100), (110) and (111) planes for the three cubic structures.

Discuss defects in crystals in relation to the electrical resistance of solids.

Explain the nature of the principal bonding mechanisms in solids and of the origin of the short-range repulsive forces.

Describe the basic steps in the preparation of integrated circuits.

Distinguish between Zener and avalanche breakdown. How do you know which mechanism is operative from the temperature dependence of the breakdown voltage of the diode?

Compare and contrast the operation and performance of field effect transistors and bipolar junction transistors.

'The junction transistor depends for its action on minority carriers.' Discuss this statement!

Describe the formation of a depletion layer when a piece of n-type semiconductor forms a junction with a piece of p-type semiconductor.

Sketch the potential energy, charge, carrier and electric field distributions across a pn junction.

What is meant by *contact potential difference*? Explain how the presence of a contact potential difference at a pn junction enables the junction to exhibit rectifying properties.

Outline the operation of the pn junction as (a) a photodetector, (b) a photovoltaic device and (c) a photoemitter.

Explain how 'pinch-off' occurs in JFETs and MOSFETs.

Explain why the collector current in a BJT increases slightly as the collector voltage increases. Discuss what the effects might be of making the collector voltage larger and larger.

Discuss the effects on the pn junction diode characteristics of (a) recombination and (b) thermal generation in the depletion layer.

Describe the operation of the phototransistor (BJT) and the optical FET (or OPFET).

Discuss which of the devices that have been described might be used in a fibre optic communications system based on silicon (SiO_2) fibre.

Answers to Problems

CHAPTER 1

1.1 4.52×10^{26} W; 1.6 kW

1.2 Setting $dI/d\lambda = 0$ gives $hc/kT\lambda_{max} = 4.965$; integrating I from 0 to ∞ gives

$$I = \frac{8\pi^5 k^4 T^4}{15 c^3 h^3}$$

1.3 228.8 nm (use Einstein's equation)

1.4 1.60×10^{-19} C (Section 1.3)

1.5 $h/(2eVm)^{1/2}$ (use de Broglie's equation)

1.6 (a) $\lambda = 6.626 \times 10^{-34}$ m (de Broglie's equation)
(b) 5×10^{-36} m (assuming you weigh 65 kg and walk at $2\,\text{m s}^{-1}$)

1.7 The resultant wave is of the form

$$y = 2A \cos(\omega t - kx) \cos(d\omega t - dkx)$$

assuming $k \gg dk$ and $\omega \gg d\omega$
This is a wave of the same form as the original constituent waves, but which has an amplitude that itself is a wave, that is $2A \cos(d\omega t - dkx)$. The velocity of this amplitude wave is $v_g = d\omega/dk$, that is the group velocity.

1.8 2.315×10^{-6} m (use the uncertainty principle)

1.9 1.054×10^{-26} J, 1.59×10^7 Hz

1.10 Think about a range of appropriate experiments that you have done and then decide.

1.11 $109.7 \times 10^5\,\text{m}^{-1}$ (use the Balmer formula)

1.12 0.136 eV (note that $E_n = E_1/n^2$)

1.13 $n = 3$ (decide which spectral series is involved)

1.14 1.05 mA (the current due to an electron moving in a circular orbit of radius r is $I = ev/2\pi r$)

CHAPTER 2

2.1 0.35 nm; 0.43 nm

2.2 Simple cubic $\pi/6$ (56%); body centred cubic $\sqrt{3}\ \pi/6$ (68%); face centred cubic $\sqrt{2}\ \pi/6$ (74%); hexagonal close packed $\sqrt{2}\ \pi/6$ (74%). *Hint:* treat the atoms as hard spheres that touch along the nearest neighbour directions then relate the atomic radius r to the lattice constant a.

2.3 (a) 1.38×10^5 (b) 8.5×10^{28} m^{-3} (c) 0.36 nm (d) 0.13 nm (e) 1.1×10^{-25} kg

2.4 0.96 nm; 0.32 nm

2.5 34.7°

2.6 0.77 nm

2.7 0.14 nm; there are no higher orders, since $n > 2$ makes $\sin\theta > 1$, which is impossible

2.8 $(\frac{1}{4}, \frac{1}{2}, 1)$; $(1, 1, \frac{1}{2})$; $(1, -1, 0)$; $(-1, \frac{1}{2}, 1)$

2.9 (a) The first term is attractive; the second is repulsive
(b) In equilibrium $dV/dr = 0$
(c) The particles will separate most easily when the force between them is a minimum, that is when $dF/dr = 0$, where $F = dV/dr$

2.10 $A \approx 10^{-37}$; $B \approx 10^{-114}$; the force to break molecule is 2.4×10^{-7} N at a separation of 0.33 nm

2.11 *Hint:* write down a series for the successive attractive and repulsive interactions

2.12 Five; grow n-type epi-layer on substrate; form oxide; open window; carry out p-type diffusion; form metal contacts

CHAPTER 3

3.1 2.54×10^{28} m^{-1} (sodium is monovalent; see Section 3.3)

3.2 1 mA

3.3 41 A (use the formula relating resistivity to sample dimensions)

3.4 $\tau = 3.03 \times 10^{-14}$ s; $\bar{v}_D = 5.3 \times 10^{-2}$ m s^{-1} (Section 3.2.1)

3.5 $\sigma = 50\,\Omega^{-1}$ m^{-1}; $\mu = 25 \times 10^{-3}$ m^2 V^{-1} s^{-1}; $\tau = 15.6 \times 10^{-14}$ s; $n = 1.3 \times 10^{22}$ m^{-3}; $v_0 = 25 \times 10^{-3}$ m s^{-1}

3.6 360 W m^{-1} K^{-2} (look up C_v in Section 1.2 and value of σ in Table 3.2)

3.7 0.84×10^{-10} m^3 C^{-1}; 7.4×10^{28} m^{-3}; 5.1×10^{-3} m^{-1} V^{-1}

3.8 10^{-7} m^2 s^{-1}; 0.18 eV (plot a graph of $\ln D$ against $(1/T)$

3.9 $4.0 \times 10^{-8}\,\Omega^{-1}$ m^{-1} (see Example 3.3)

CHAPTER 4

4.1 Substitute the second partial derivatives of the given function into a two-dimensional Schrödinger equation, collect terms and solve for E.

4.2 1.4×10^{-18} J

4.3 137.4 nm (use Planck's formula)

4.4 Use equation (4.14) with limits 0 and d to show that the amplitude of the wave function is $\sqrt{2/d}$.

4.5   1.97 eV (integrate $n(E)dE$ between 0 and E_F; read Section 4.9)
4.6 7.4×10^5 m s^{-1} (see Section 3.2.1)
4.7   All the zones have area $(\pi/a)^2$

CHAPTER 5

5.1 $2.23\,\Omega^{-1}\,m^{-1}$; $1.43 \times 10^{-6}\,\Omega^{-1}\,m^{-1}$ (use equation 5.3)
5.2 $E_d = 7.7 \times 10^{-3}$ eV; $E_a = 5.7 \times 10^{-2}$ eV (use equation 5.6 with appropriate effective masses)
5.3 $n_i = 6.2 \times 10^{21}$ m^{-3} ($\sigma = 1/\rho$, then use equation 5.3)
5.4 $E_g = 1.119$ eV (the conductivity is proportional to $\exp(-E_g/2kT)$
5.5 $n_i = 2.42 \times 10^{19}$ m^{-3} (evaluate N_c and $\exp(-E_g/2kT)$)
5.6 $E_g = 0.67$ eV (plot graph of ln σ versus $1/T$)
       $\lambda_g = 1.85 \times 10^{-6}$ m (use $\lambda_g = hc/E_g$)
5.7 $p = 1.64 \times 10^{22}$ m^{-3}, $\mu_h = 0.041$ m^2 V^{-1} s^{-1} (the Hall coefficient is positive; therefore holes are the majority carriers and hence we can assume $\sigma = pe\mu_h$)
5.8   (a) $n = 1 \times 10^{24}$ m^{-3} (donors, thus $n \approx N_d$ and $np = n_i^2$)
              $p = 1 \times 10^2$ m^{-3}, $\sigma_n = 1.36 \times 10^5\,\Omega^{-1}\,m^{-1}$
       (b) $n = 1 \times 10^{24}$ m^{-3}
              $p = 5.76 \times 10^{14}$ m^{-3}, $\sigma_n = 6.25 \times 10^4\,\Omega^{-1}\,m^{-1}$
5.9   (a) $p = 3.28 \times 10^{21}$ m^{-3}; $n = 1.76 \times 10^{17}$ m^{-3}
       (b) $p = 1.3 \times 10^{22}$ m^{-3}; $n = 1.73 \times 10^{10}$ m^{-3}
5.10 $\mu = 0.4$ m^2 V^{-1} s^{-1}
5.11 $J = 1.125 \times 10^6$ A m^{-2}, $I = 1.125$ A (use equation 5.20a)
5.12 $\tau = 2.47\,\mu s$ (use equation 5.23)
5.13 $E_g = 2.455$ eV, $E_d = 6.2 \times 10^{-2}$ eV (must be n-type as the Hall coefficient is negative)

CHAPTER 6

6.1   For an ideal junction, where $\phi_m > \phi_s$ an ohmic junction is formed, while for $\phi_m < \phi_s$ a rectifying junction is formed.
6.2 $V_0 = 0.75$ V, $W = 2.74 \times 10^{-6}$ m (use equations 6.6 and 6.7), also $x_n = 0.999$ W and $x_p = 0.001$ W
6.3   (a) $I = 2.91 \times 10^{-5}$ A, $r_e = 859\,\Omega$ (use equations 6.16 and 6.19)
       (b) $I = 0.07$ A, $r_e = 0.36\,\Omega$
6.4 $C_0 = 52.7$ pF, $C_5 = 19.1$ pF (use equations 6.17 and 6.7)
6.5   The energy gained by the accelerated electron in a distance equal to the mean free path must equal E_g, that is $(eV_{br}/W)\lambda = E_g$ (assuming a

uniform field) and E_{br} across the depletion layer. Substituting for W gives $V_{br} = 40.92$ V

6.6 $\beta = 437$, $\alpha = 0.9977$

6.7 $V_p = 5.44$ V

6.8   For light emission the material must have a direct band gap; hence (a) silicon, (b) gallium arsenide, (c) gallium phosphide, (d) silicon carbide

6.9   3.77 mA

6.10   11.29 mA; 0.348 V

6.11   Sensitivity $= 8.86 \times 10^{-5}$ W m^{-2}, corresponding to 7.8×10^9 photons per second. The responsivity of the photodiodes is expressed in amperes per watt, that is 0.45 A W^{-1}

Appendix 1
Properties of Semiconducting Materials (at 300 K)

Semiconductor	Energy gap E_g (eV)	Electron mobility μ_e (m²V⁻¹s⁻¹)	Hole mobility μ_h (m²V⁻¹s⁻¹)	Intrinsic carrier concentration n_i (m⁻³)	Conductivity[a] ρ (Ω⁻¹ m⁻¹)	Electron effective mass m_e^* (μm)	Hole effective mass m_h^* (μm)	Relative permittivity ϵ_r	Lattice constant a (nm)	Melting point (°C)
Si (i)[b]	1.12	0.135	0.048	1.5×10^{16}	4.4×10^{-4}	0.26	0.38	11.8	0.543	1415
Ge (i)	0.67	0.39	0.19	2.4×10^{19}	2.2	0.12	0.23	16.1	0.566	937
GaAs (d)	1.43	0.85	0.04	1.0×10^{13}	2.5×10^{-7}	0.067	0.5	10.9	0.565	1237
GaP (i)	2.25	0.03	0.015	1.4×10^{5}	9.9×10^{-1}	0.12	0.67	8.5	0.545	1467
InP (d)	1.34	0.46	0.015	1.2×10^{14}	1.25×10^{6}	0.07	0.6	12.4	0.587	1070
InAs (d)	0.36	3.3	0.045	6.0×10^{21}	3.3×10^{3}	0.022	0.4	14.6	0.506	943
InSb (d)	0.18	8.0	0.045	1.0×10^{22}	1.67×10^{3}	0.013	0.18	16.7	0.648	525
CdS (d)	2.5	0.03	0.002			0.21	0.8	7.1	0.414	1475
PbS (i)	0.37	0.06	0.02		2.0×10^{6}	0.25	0.25	161	0.594	1119
PbSe (i)	0.26	0.12	0.12		1.0×10^{5}	0.33	0.34	280	0.615	1081
ZnS (d)	3.6	0.01	—		1.0×10^{-8}	0.4	—	7.0	0.541	1650
ZnTe (d)	2.3	0.034	0.01		1.0	—	—	10.4	0.61	1238 (Vaporizes)
SiC (d)	2.86	0.09	0.005			0.85	1.0	7.0	1.511	2830

[a]The conductivity is for high purity material — for Si, Ge, GaAs and GaP it has been extrapolated to intrinsic material. The data have been taken from a number of references. The values given often do not agree closely; this applies particularly to the effective masses and relative permittivity.

[b]The symbols (i) or (d) indicate indirect or direct energy bandgap.

Appendix 2
SI Units and Values of Physical Quantities

DERIVED SI UNITS

Name of quantity	Symbol	SI unit
Mass	m	kg
Force	F	$\text{kg m s}^{-2} = \text{N (newton)}$
Momentum	p	$\text{kg m s}^{-1} = \text{N s}$
Pressure	$p \ldots P$	$\text{N m}^{-2} = \text{Pa (pascal)}$
Work	$w \ldots W$	$\text{kg m}^2 \text{s}^{-2} = \text{J (joule)}$
Energy	E	J*
Power	P	$\text{J s}^{-1} = \text{W (watt)}$
Thermodynamic temperature	$T \ldots \theta$	K (kelvin)
Thermal conductivity	$\lambda \ldots \varkappa$	$\text{J s}^{-1} \text{m}^{-1} \text{K}^{-1} = \text{W m}^{-1} \text{K}^{-1}$
Electric current	I	A (ampere)
Electric charge	Q	$\text{A s} = \text{C (coulomb)}$
Electric potential	V	$\text{kg m}^2 \text{s}^{-3} \text{A}^{-1} = \text{V (volt)}$
Electric potential difference	$U \ldots V$	V
Magnetic induction	B	$\text{V s m}^{-2} = \text{Wb m}^{-2} = \text{T (tesla)}$
Magnetic flux	ϕ	$\text{V s} = \text{Wb (weber)}$
Magnetic field strength	H	A m^{-1}
Magnetic susceptibility	χ_m	Dimensionless
Self-inductance	L	$\text{Wb A}^{-1} = \text{H (henry)}$
Permeability	μ	H m^{-1}
Magnetization	M	A m^{-1}
Resistance	R	$\text{V A}^{-1} = \Omega$
Resistivity	ρ	$\Omega \text{ m}$
Conductance	G	$\Omega^{-1} = \text{S (siemens)}$
Conductivity	$K \ldots \sigma$	$\Omega^{-1} \text{m}^{-1}$
(Molar) gas constant	R	$\text{J K}^{-1} \text{mol}^{-1}$
Molar heat capacity	C_v, C_p	$\text{J K}^{-1} \text{mol}^{-1}$

*(In many books on solid state physics and related topics energy is expressed in electron volts (eV), though this is not strictly a unit but the product of the electron charge e and a voltage V.)

255

RECOMMENDED VALUES OF
SOME PHYSICAL QUANTITIES

Name of quantity	Symbol	Value
Speed of light	c	$(2.997925 \pm 0.000003) \times 10^8 \text{ m s}^{-1}$
Permeability of vacuum	μ_0	$4\pi \times 10^{-7} \text{ kg m s}^{-2} \text{A}^{-2}$ (exactly)
Permittivity of vacuum	ϵ_0	$(8.854185 \pm 0.000018) \times 10^{-12}$ $\text{kg}^{-1} \text{m}^{-3} \text{s}^4 \text{A}^2$
Mass of electron	m	$(9.1091 \pm 0.0004) \times 10^{-31} \text{ kg}$
Charge of electron	e	$(1.60210 \pm 0.00007) \times 10^{-19} \text{ C}$
Mass of proton	M_H	$(1.67252 \pm 0.00008) \times 10^{-27} \text{ kg}$
Boltzmann's constant	k	$(1.38054 \pm 0.00018) \times 10^{-23} \text{ J K}^{-1}$
Planck's constant	h	$(6.6256 \pm 0.0005) \times 10^{-34} \text{ J s}$
Rydberg constant for hydrogen	R_H	$(1.0967758 \pm 0.0000003) \times 10^7 \text{ m}^{-1}$
Bohr magneton	μ_B	$(9.2732 \pm 0.0006) \times 10^{-24} \text{ A m}^2$
Avogadro's constant	L, N	$(6.02252 \pm 0.00028) \times 10^{23} \text{ mol}^{-1}$
Stefan constant	σ	$5.6697 \times 10^{-8} \text{ W m}^{-2} \text{K}^{-4}$

Appendix 3
Motion of an Electron in a Periodic Lattice: the Kronig–Penney Model

If the Schrödinger equation is applied to the problem of electron motion through the periodic array of atoms in crystals, the existence of energy bands is predicted. The simple model of a one-dimensional array of atoms proposed by Kronig and Penney will now be considered.

Figure A3.1. The potential energy of an electron in a linear periodic lattice of positive nuclei

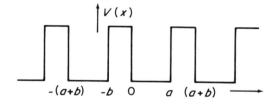

Figure A3.2. The potential distribution in a one-dimensional square-well lattice

The potential energy of an electron moving in the neighbourhood of the nuclei is shown in Figure A3.1, and the square-well approximation to this situation used by Kronig and Penney is shown in Figure A3.2. The time-independent Schrödinger equation takes the following form in the various regions: for $0 < x < a$,

$$\frac{d^2 \psi}{dx^2} + \frac{8\pi^2 m}{h^2} E \psi = 0 \tag{A3.1}$$

and for $-b < x < 0$,

$$\frac{d^2\psi}{dx^2} + \frac{8\pi^2 m}{h^2}(E - V_0)\psi = 0 \qquad (A3.2)$$

Let

$$\alpha^2 = \frac{8\pi^2 mE}{h^2}$$

and

$$\beta^2 = \frac{8\pi^2 m}{h^2}(E - V_0)$$

A solution must be found that will be appropriate for both of these regions and that will satisfy the boundary conditions at the potential barriers. Bloch showed that the solution is described by a plane wave whose amplitude is modulated by the periodicity $a + b$ of the lattice. The electron is thus described by a wave ψ with a wave number $k = 2\pi/\lambda$ and a periodically varying amplitude; that is

$$\psi = e^{ikx}u(x)$$

where e^{ikx} represents the oscillating part of the plane wave and $u(x)$ represents the modulation. Substituting this into equations (A3.1) and (A3.2) gives, for $0 < x < a$,

$$\frac{d^2 u}{dx^2} + 2ik\frac{du}{dx} - (k^2 - \alpha^2)u = 0 \qquad (A3.3)$$

and for $-b < x < 0$,

$$\frac{d^2 u}{dx^2} + 2ik\frac{du}{dx} - (k^2 + \beta^2)u = 0 \qquad (A3.4)$$

It can be verified by substitution that solutions for these two equations may be written as: for $0 < x < a$,

$$u = A\, e^{i(\alpha - k)x} + B\, e^{-i(\alpha + k)x} \qquad (A3.5)$$

and for $-b < x < 0$,

$$u = C\, e^{(\beta - ik)x} + D\, e^{-(\beta + ik)x} \qquad (A3.6)$$

Equations (A3.5) and (A3.6) represent two waves travelling in the positive and negative x directions respectively. All four waves are subject to the boundary conditions that ψ and $d\psi/dx$ (that is u and du/dx) must be continuous at $x = 0$ and $x = a$, and that u should be periodic that is $u(x = -b) = u(x = a) = \ldots$.

Applying these boundary conditions, four equations are obtained that ensure that the values of A, B, C and D are such that equations (A3.3) and (A3.4) will be solutions of the Schrödinger wave equation. These four restrictive equations are as follows. For continuity of u at $x = 0$,

$$A + B = C + D$$

For continuity of du/dx at $x = 0$,

$$Ai(\alpha - k) - Bi(\alpha + k) = C(\beta - ik) - D(\beta + ik)$$

For periodicity of u,

$$A\, e^{i(\alpha - k)a} + B\, e^{-i(\alpha + k)a} = C\, e^{(\beta - ik)(-b)} + D\, e^{(\beta + ik)(-b)}$$

For periodicity of du/dx,

$$i(\alpha - k)A\, e^{i(\alpha - k)a} - i(\alpha + k)B\, e^{-i(\alpha + k)a} = (\beta - ik)C\, e^{(\beta - ik)(-b)}$$
$$- (\beta + ik)D\, e^{-(\beta + ik)(-b)}$$

For these equations to be consistent, the determinant of the coefficients of A, B, C and D must be zero. This leads to

$$\frac{\beta^2 - \alpha^2}{2\alpha\beta} \sinh \beta b \, \sin \alpha a + \cosh \beta b \, \cos \alpha a = \cos k(a + b) \qquad \text{(A3.7)}$$

This rather difficult equation can be simplified if b is allowed to decrease and at the same time V_0 to increase, such that the product $V_0 b$ remains finite. The model is thus modified to one of a series of wells separated by infinitely thin potential barriers. With this modification, equation (A3.7) becomes

$$\frac{1}{2\alpha} \frac{8\pi^2 m(V_0 b)}{h^2} \sin \alpha a + \cos \alpha a = \cos ka \qquad \text{(A3.8)}$$

or

$$\frac{P \sin \alpha a}{\alpha a} + \cos \alpha a = \cos ka \qquad \text{(A3.9)}$$

where

$$P = \frac{4\pi^2 m a(V_0 b)}{h^2}$$

Equation (A3.9) gives values of α that permit solutions of the electron wave equation to exist. Now α is a function of the electron energy E so once again the quantum mechanical treatment of electrons restricts the energy to certain allowed values, the restriction depending on P, which in turn depends on the potential barrier strength $V_0 b$.

In equation (A3.9), the left-hand side (following the right-hand side) is restricted to values between ± 1. Figure A3.3 shows a plot of $(P \sin \alpha a)/\alpha a + \cos \alpha a$ against αa for P equal to $3\pi/2$. The regions in which allowed solutions exist are marked by heavy lines. Thus the motion of an electron in a periodic lattice is characterized by bands of allowed energy separated by forbidden regions.

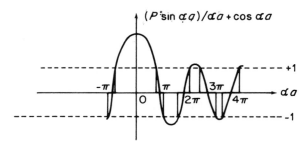

Figure A3.3. Plot of $(P \sin \alpha a)/\alpha a + \cos \alpha a$ against αa for $P = 3\pi/2$

Index

Absorption,
  edge, 169
  spectra, 171
Acceptor impurities, 161
Aggregates of atoms, 128
Alloys,
  resistivity of, 91
Arrays,
  disordered, 33
  ordered, 33
Atomic bonds, 51
  covalent, 55
  ionic, 53
  metallic, 57
  van der Waals, 57
Atomic
  radius, 48
  structure, 20
Atomic theory of magnetism, 94
Avalanche breakdown, 206

Balmer formula, 20
Base transit time, 215
Black body radiation, 2
Bloch function, 258
Bloch wall, 105
Bohr,
  magneton, 95
  postulates, 23
Bohr theory, 22
Boltzmann distribution, 142
Bonds,
  covalent, 55
  ionic, 53
  metallic, 57
  van der Waals, 57
Bragg law, 37
Bravais lattice, 44
Brillouin zones, 133
Bubble raft, 62

Carriers,
  lifetime, 177

majority, 172
minority, 172
Classical conduction,
  breakdown of, 87
  model of, 81
Classification of solids, 145
Colour centres, 59
Common base operation, 214
Common emitter operation, 218
Conductivity,
  electrical, 84
  extrinsic, 166
  intrinsic, 166
  ionic, 81
  semiconductors, 165
  thermal, 86
Contact potentials,
  measurement of, 185
  table of, 186
Cooper pairs, 112
Crystal axes, 41
Crystal defects, 58
Crystal geometry, 39
Crystal growth, 66
  Bridgeman–Stockbarger method, 66
  Czochralski method, 66
  floating-zone method, 68
Crystal symmetry elements, 39
Crystal lattice, 42
Crystal planes, 41
Crystal structures,
  cubic, 47
  hexagonal, 48
  hexagonal close packed, 49
  spinel, 108
Crystal systems, 45
Curie law, 102
Curie temperature, 104
Curie–Weiss law, 105

Davisson and Germer experiment, 13
de Broglie hypothesis, 13
Debye, 7

Degeneracy, 125
    removal of, 127
Density of states function, 143
Depletion layer,
    p–n junction, 194
    width, 195
Device fabrication, 68
    field effect transistors, 75
    impurity diffusion, 70
    ion implantation, 71
    junction transistors, 74
    window opening, 71
Diffusion,
    constant, 174
    current, 174, 197
    potential, 196
Diode,
    laser, 234
    light emitting, 231
    p–n junction, 194
    tunnel, 207
    Zener, 204
Dislocations,
    edge, 59
    screw, 61
Distribution,
    Fermi–Dirac, 144
    Maxwell–Boltzmann, 142
Domains,
    ferromagnetic, 103
    movement of, 105
Donor levels, 160
Doping, 159
Drift current, 197
Drift velocity, 83
Drude theory, 82

Effective mass, 135
Eigenvalues, 119
E–k curves, 133
Einstein, 2, 12, 15
Einstein relations, 174
Electron,
    configuration, 29
    diffraction, 13
    gas, 81
    probability distribution, 144
    shells, 29
    spin, 96
Energy bands,
    conduction, 131

experimental investigation, 138
    filled, 131
    in solids, 128
    overlap, 145
    partially filled, 144
    p-band, 130
    s-band, 130
    valence, 131
Energy gap, 131
Energy level diagram, 25
Epitaxial growth, 68
Equipartition of energy, 4
Etch pits, 63
Exchange integral, 104
Excitation energy of impurities, 160

Fermi–Dirac distribution, 144
Fermi function, 143
Fermi level,
    effect of doping on, 164
    equalization of, 184
    in semiconductors, 162
    table of, 140
    temperature variation of, 165
Ferrimagnetism, 108
Ferrites, 110
Ferromagnetism, 102
Field-effect transistor, 220
Frenkel defect, 59

Hall coefficient, 90
Hall effect, 88
Hall voltage, 89
Haynes–Shockley experiment, 172
Heisenberg Uncertainty Principle, 17
Holes, 151
Hysteresis, 103

Intensity of magnetization, 99
Interstitial impurity, 59
Intramolecular field, 104
Inversion layer, 193, 224
Ionization energy, 24

Junction
    capacitance, 210
    metal–metal, 183
    metal–semiconductor, 189
    semiconductor–semiconductor, 194
    slope resistance, 211

Junction transistor,
  bipolar, 212
    characteristics, 216
    npn, 214
    pnp, 215
  field effect, 220
    JFET, 221
    MOSFET, 223
    characteristics, 223, 225

Kronig–Penney model, 131

Lasers
  doped insulator, 236
  fundamentals of operation, 233
  pn-junction, 234
Lattice constant, 48
Lattice vibrations, 6, 63
Light emitting diodes, 231

Magnetic moment,
  orbital, 94
  spin, 96
Magnetic properties of solids, 92
Magnetic quantum numbers, 96
Magnetic susceptibility, 101
Magnetism,
  classification of, 99
Mathiessen's law 91
Majority carriers, 160
Mean free paths, 82, 87
Miller indices, 46
Minority carrier,
  dynamics, 172
  lifetime, 177
  mobility, 173
Mobility,
  temperature variation, 166

Nearest neighbour distance, 48

Ohmic contacts, 193
Ohm's law, 83
Orbitals,
  p-, 127
  s-, 127

Paramagnetism, 102
Particle in a box, 121
  one dimension, 121
  three dimensions, 124

Pauli exclusion principle, 28
Peltier effect, 188
Permeability, 101
Phonons, 64
Photoconductivity, 226
Photodiodes, 228
Photoelectric effect, 8
Pinch-off voltage, 223
Planck's constant, 5
Plank's hypothesis, 5
Plane defects,
  grain boundary, 62
  stacking fault, 62
p–n junction diode,
  characteristics, 202
    forward biased, 201
    reverse biased, 199
    temperature effects on, 203
Point defects, 58
Probability density, 120

Quantum numbers, 28
Quantum theory of conduction, 147

Rayleigh–Jeans law, 4
Recombination,
  direct 175
  indirect, 176
Rectifiers,
  p–n junction, 202
  Schottky, 192
Reduced Zones 134
Reverse bias breakdown, 203
Reverse saturation current, 201
Richardson–Dushman equation, 180
Rutherford scattering, 21
Rydberg constant, 20

Saturation magnetization, 108
Shockley diode equation, 201
Schrödinger equation,
  time dependent, 118
  time independent, 119
Seebeck effect, 186
Semiconductor,
  diode laser, 233
  direct bandgap, 176
  extrinsic, 159
  indirect bandgap, 176
  intrinsic, 158
  optical absorption in, 171

Semiconductor (*cont.*)
  optical properties of, 169
Solar cell, 230
Specific heat, 6
Spectral lines, 19
Spectral series, 20
Spin quantization, 98
Stark effect, 128
Statistics,
  classical, 142
  Fermi–Dirac, 143
  Maxwell–Boltzmann, 142
  quantum, 143
Stopping potential, 10
Substitutional impurity, 59
Superconductivity, 111
Surface states, 192
Susceptibility, 101

Thermal velocity, 84
Thermionic emission, 178
Thomson and Reid experiment, 15
Thomson effect, 188
Transistor,
  amplifier, 218
  switching mode, 219
Traps, 59, 175
Tunnel diode, 208
Tunnelling, 207

Ultraviolet catastrophe, 5
Unit cell,
  crystal, 43

Vacancy, 58
Vacuum level, 169, 179
Valence band, 131
Valence electrons, 29
Vector model of the atom, 28

Wave equation, 117
Wave function,
  physical interpretation of, 119
Waves packets, 17, 134
Wave-particle duality, 13
Wiedemann–Franz,
  constant, 87
  law, 85
Wien displacement law, 3
Work function, 11

X-ray diffraction,
  powder method, 38
X-ray spectra,
  soft, 138

Zeeman effect, 128
Zener breakdown, 204
Zener diode, 207
Zone refining, 65